Nanoscale Science and Engineering Forum

AA009251

Held at the 2023 AIChE Annual Meeting

Orlando, Florida, USA
5-10 November 2023

ISBN: 978-1-7138-9310-3

Printed from e-media with permission by:

Curran Associates, Inc.
57 Morehouse Lane
Red Hook, NY 12571

Some format issues inherent in the e-media version may also appear in this print version.

Copyright© (2023) by AIChE
All rights reserved.

Printed with permission by Curran Associates, Inc. (2024)

For permission requests, please contact AIChE
at the address below.

AIChE
120 Wall Street, FL 23
New York, NY 10005-4020

Phone: (800) 242-4363
Fax: (203) 775-5177

www.aiche.org

Additional copies of this publication are available from:

Curran Associates, Inc.
57 Morehouse Lane
Red Hook, NY 12571 USA
Phone: 845-758-0400
Fax: 845-758-2633
Email: curran@proceedings.com
Web: www.proceedings.com

TABLE OF CONTENTS

AREA PLENARY: BIONANOTECHNOLOGY (INVITED TALKS)

3a Micro/Nanoengineered Adhesive Biomaterials for Regenerative Medicine and Drug Delivery 1
Nasim Annabi

3b Harnessing Chaos: Rational Design of Nanoscale Surfaces for Cavitation-Based Imaging Agents, Nanomotors, and Protein Therapeutics. 2
Andrew Goodwin

3c Building Bioactivity and Nanostructure into Slippery Liquid-Infused Porous Surfaces 3
David M. Lynn

AREA PLENARY: CARBON NANOMATERIALS (INVITED TALKS)

37a Exploring the Best CVD Conditions for Growth of Small-Diameter Single-Wall Carbon Nanotubes Using an Autonomous Research System 4
Brian M. Everhart, Rahul Rao, Tsung Wei Liu, Diego Gomez Gualdron, Benji Maruyama, Placidus Amama

37b Molecular Design and Investigation of 2-D Hybrid Materials 5
Sanket Deshmukh

BIONANOTECHNOLOGY GRADUATE STUDENT AWARD SESSION

370a Award Submission: Tracking Dendritic Cell Vaccination Using Magnetic Particle Imaging 6
Bo Yu

370b Award Submission: Accelerated Blood Clearance of Lipid Nanoparticles Occurs Regardless of Formulation 7
Mariah Arral

370c Award Submission: Mucus Penetration of Surface-Engineered Nanoparticles in Various pH Microenvironments 8
Yiyang Guo

370d Award Submission: On-Demand Drug Release of Polymeric Core-Shell Nanoparticles Developed through Aqueous Surface-Initiated Sara-ATRP 10
Nahieli Preciado

370e Award Submission: Detection and Visualization of HER2 Clustering Using an Aggregation Induced Emission Based Fluorescent Probe Molecule 13
Yasaman Moradi

370f Award Submission: Covalent Attachment of Horseradish Peroxidase to Single-Walled Carbon Nanotubes for Hydrogen Peroxide Detection 14
Francis Ledesma

370g Award Submission: Desirable Formulations for Inhalable Particulate Immunotherapeutics Using Model Peg-Based Nanoparticle 15
Emma Sudduth, Catherine Fromen

370h Using Nanoparticle-Gated Electrokinetic Membrane Sensors and Janus Diffusivity Sensors for Quantifying Disease-Related Nanovesicles in Biofluids 16
Sonu Kumar, John Sinclair, Nalin Maniya, Satyajyoti Senapati, Hsueh-Chia Chang

CARBON NANOMATERIALS I: DISPERSION, SURFACE STRUCTURE, AND BIOINTERACTIONS

220a Investigating Ergodicity-Broken Rotational Dynamics of SWCNT in Hexagonally Packed Colloidal Pores Via Machine Learning 18
Zhao Tang, Shannon Eichmann, Vida Jamali, Fred Mackintosh, Matteo Pasquali

220b Wetting Transparency of Free-Suspended Single-Layer Graphene on Liquid Substrate 19
Fan Yang, Annette Thompson, Lei Li

220c Green Fluorescent Carbon Dots for Sensitive and Selective Detection of Ferrous Ions 20
Xiangcheng Sun

220d Quantified Carbon Nanotube (n,m) Separation Dependencies in Aqueous Two-Polymer Phase Extraction 21
Christopher Sims, Jeffrey Fagan

220e Catalyst Aerosol Dynamics in the Floating Catalyst Chemical Vapor Deposition Synthesis of High-Quality Carbon Nanotubes 22
Arthur Sloan, Muxiao Li, Jui Junnarkar, Glen Irvin, Matteo Pasquali

220f Macroscopic Weavable Fibers of Carbon Nanotubes with Giant Thermoelectric Power Factor 24
Natsumi Komatsu, Yota Ichinose, Oliver Dewey, Lauren Taylor, Mitchell Trafford, Yohei Yomogida, Geoff Wehmeyer, Matteo Pasquali, Kazuhiro Yanagi, Junichiro Kono

220g Thermal Conductivity of Interlayer-Bonded Bilayer Graphene 25
Afnan Mostafa, Ashwin Ramasubramaniam, Dimitrios Maroudas

CARBON NANOMATERIALS: GRADUATE STUDENT AWARD SESSION

281a Controlled Synthesis of Carbon Nanoparticles from Lignin Using Single-Step Aerosol Methods for Application in Energy Storage 26
Sujit Modi, Onochie Okonkwo, Sulay Saha, Hao Zhou, Shalinee Kavadiya, Marcus Foston, Pratim Biswas

281b Obtaining Structural Information of Carbon Black in Complex Mediums for Energy Storage Systems with Rheo-Electric Measurements 28
Qingsong Liu, Yoshita Gupta, Jeffrey Richards

281c Investigation of the Electrical Properties of Carbon Black Suspensions Undergoing Shear-Induced Microstructural Rearrangement 29
Paolo Ramos, Julie Hipp, Qingsong Liu, Jeffrey Richards

281d Polarization Resolved Fluorescence Detection of Carbon Nanotube Corona Phases 30
Xiaojia Jin, Matthias Kuehne, Xun Gong, Amandine Ucko, Volodymyr Koman, Michael Strano

NANOMATERIALS FOR ENERGY CONVERSION

244a Investigation of Advanced Catalysts and Metal Oxide Supports Having Nanoporous Structures Towards Oxygen Evolution Reaction in Proton Exchange Membrane Water Electrolyzer.............. 31
Chaekyung Baik, Je Yeon Choi, Hyung Joo Lee, Chanho Pak

244b Mechanistic Role of Cations in the Electric Double Layer during Electrocatalysis................................ 32
Hyungjun Kim

244c Facile Defect Engineering to Tailor Porous Coordination Polymers for Energy-Efficient Gas Separation.. 33
Jong Suk Lee

244d Nanophase-Segregated Structures and Transport in Aquivion Membrane for Polymer Electrolyte Membrane Fuel Cell: Molecular Modeling Approach 34
Mohammed Bazaid, Seung Soon Jang

244e First-Principles Investigation of Band-Gap Tuning in 2D Hybrid Organic-Inorganic Perovskites.. 35
Omar Allam, Yoonseo Nah, Ilgeum Lee, Dong Ha Kim, Seung Soon Jang

244g Photothermal Effects of Plasmonic Nanoparticles to Enhance Diels–Alder (DA) Reactions in Reversible Thermosets 36
Arnob Dipta Saha, Brandon McReynolds, Madeline Finale, Youngmin Lee, John McCoy, Sanchari Chowdhury

244h Thermoelectric Transport Coefficients for Weyl Semimetals in the Family of Transition Metal Monopnictides with Dislocation Deffects 37
Enrique Munoz

NANOMATERIALS FOR ENERGY STORAGE

300a Molecular Design of Polymer for Solid State Electrolyte Application in Li-Ion Battery: Molecular Modeling Approach... 39
Omar Allam, Seung Soon Jang

300b Unravelling Ultrafast Li-Ion Transport in Functionalized Metal-Organic-Framework Based Battery Electrolytes 40
Zheng Chen

300c Computational Design of High Performance Liquid Organic Hydrogen Carriers and Their Corresponding Dehydrogenation Catalysts 41
Jeong Woo Han

300d Operando Study Electronic and Magnetic Transitions in Lithium and Sodium Batteries Using X-Ray Emission/Absorption Spectroscopy 42
Faisal M. Alamgir

300e Developing Non-Aqueous and Low-Temperature Iron Batteries with V_2O_5 Nanowire Cathodes.. 43
Raju Vadthya, Hao Nguyen, Matt Powell, Shuya Wei

300f Conductive Membrane Coating for High Performance Vanadium Redox Flow Battery Achieved By Air-Controlled Electrospray.. 45
Yiqi Shao, Xiaosi Gao, Shuo Jin, Yong Lak Joo

300g Localized High Concentration Electrolytes for Dual Ion Lithium Batteries 46
Youngbi Kim, Jeong Woo Han

300h Engineering Block Copolymer-Derived Porous Carbon Nanofibers As Anodes for Sodium-Ion Batteries 47
Kenneth Gordon, Cameron Romero, Zhi Liu, Karius Joseph, Emily Broussard, Griffin Usie, Ling Fei

MACHINE LEARNING FOR NANOMATERIALS FOR ENERGY APPLICATIONS

159a Machine Learning Technique for Core Materials Design Toward Active Systems in Energy Storage and Conversion Reactions 48
Byungchan Han, Hoje Chun, Minjoon Hong

159b Modeling and Design of Single-Atom Alloy Catalysts for CO_2 Hydrogenation Reactions 49
Mie Andersen, Raffaele Cheula

159c Machine-Learning Driven Exploration of Catalytic Reaction Network 50
Hyunwook Jung, Sina Stocker, Karsten Reuter, Johannes T. Margraf

159d Generative AI for Materials Modeling 51
Fei Zhou

159e Accelerating the Discovery of Organic Electrode Materials for Li-Ion Batteries: An Integrated Approach Using DFT, Machine Learning, and High-Throughput Virtual Screening 52
Omar Allam, Seung Soon Jang

159f Active Learning Exploration of Single Atom Active Sites for Oxygen Reduction and Evolution Reactions 53
Hoje Chun, Rafael Gomez-Bombarelli, Byungchan Han

BIONANOTECHNOLOGY FOR GENE AND DRUG DELIVERY

141a Stepwise Bulk Process for Enhanced Productivity and Stability of mRNA Vaccines 54
Minsub Chung, Sangmin Lee, Hyun Jin Kim

141b Lipid Nanoparticle Permeabilization and Fusion Quantified By Two High Throughput Fluorescence Assays 56
Alison Lui, Mariah Arral, Elizabeth Voke, Kathryn Whitehead, Kranthi K. Mandadapu, Markita Landry

141c Nanoparticle Delivery Doubles the Antinociception Effect of a CGRP Receptor Antagonist in Oral Cancer Pain Models 57
Parker Lewis, Tu Huu Nguyen, Varun Chokshi, Rachel Pollard, Nigel Bunnett, Brian Schmidt, Nathalie Pinkerton

141d Block-Copolymer Stabilized Lipid Nanoparticles: A Novel Platform for Targeted Gene Delivery 59
Sai Nikhil Subraveti, Bumjun Kim, Robert K. Prud'Homme

141e Rational Design of Silica-Based Nanoparticles for Overcoming Barriers of siRNA Delivery in Relapsed Prostate Cancer Applications 60
Angelea Maestas-Olguin

141f Influence of Protein Corona on Lipid Nanoparticle Uptake 61
Elizabeth Voke, Mariah Arral, Rebecca L. Pinals, Henry Squire, Alison Lui, Anthony Lavarone, Kathryn Whitehead, Markita Landry

141g Reactive Oxygen Species (ROS)-Responsive Doxorubicin Prodrug for Breast Cancer Treatment 62
Mina Jafari, Sophia Schutte, Lisa Privette Vinnedge, Joo-Youp Lee

NANOSCALE SCIENCE AND ENGINEERING FORUM DIVISION PLENARY

185a NSEF Young Investigator Award Winner: Nanoscale science and Engineering of Atomically Thin Membranes for Next-generation Healthcare, Energy, & Environmental Applications 63
Piran Kidambi

185b NSEF Forum Award Winner: Making optical materials from metal oxide nanocrystals 64
Delia Milliron

BIONANOTECHNOLOGY FOR SENSORS AND IMAGING

299a Invited Talk: Carbon Nanotube Quantum Well Defect Emission for Machine Learning-Guided Diagnostics 65
Daniel Heller, Mijin Kim, Anand Jagota, Yuhuang Wang, Ming Zheng

299b Hardware and Sensor Optimization of a Nanotechnology Enabled Implantable Sensor for Hormones in Marine Animals 66
Volodymyr Koman, Xiaojia Jin, Manki Son, Michael Strano

299c Covalent Attachment of Horseradish Peroxidase to Single-Walled Carbon Nanotubes for Hydrogen Peroxide Detection 67
Francis Ledesma, Shoichi Nishitani, Francis J. Cunningham, Linda Chio, Aishwarya Murali, Markita Landry

299e High-Throughput Evolution of Near-Infrared Oxytocin Nanosensors 68
Jaquesta Adams, Nicole Navarro, Esther Leem, Xiaoqi Sun, Tavio Arias, Markita Landry

299f Glycopolymer-Wrapped Carbon Nanotubes for Sars-Cov-2 Spike Protein Detection 69
Brandon Heppe, Geyou Ao, Xue-Long Sun

299g Decoding Early Stress Signaling Waves in Plants Using Multiplexed Nanosensors 70
Mervin Ang, Jolly Saju, Thomas K. Porter, Sarangapani Sreelatha, Duc-Thinh Khong, Suh In Loh, Song Wang, Gajendra Pratap Singh, Nam-Hai Chua, Michael Strano, Rajani Sarojam

299h Imaging Dopamine Modulation in Brain Acute Slices for Cocaine Addiction Study with Non-Genetically Encoded Nanosensors 71
Nicholas Ouassil, Natsumi Komatsu, Markita Landry

299i Wide-Range Size Tuning of Conjugated Polymer Nanoparticles for Multi-Color Bioimaging 72
Noriko Nakamura, Mikito Tokumaru, Seiichi Ohta

PLENARY SESSION FOR NANOMATERIALS FOR ENERGY APPLICATIONS (INVITED TALKS)

51a Direct Air Capture (DAC) of CO_2 Using Porous Sorbent Materials Coupled with Scalable Processes 75
Christopher Jones

51b Thermally Stable Single Atom Catalysts for Energy Conversion 76
Abhaya Datye

51c Computational Materials Design with Machine Learning and Atomistic Simulations 77
Rafael Gomez-Bombarelli

POSTER SESSION: NANOSCALE SCIENCE AND ENGINEERING FORUM

30a Electrochemical Windows of Sulfone-Based Electrolytes for Lithium Metal Batteries: A Density Functional Theory and Cluster-Continuum Model Investigation 78
Minhee Park, Junbeom Cho, Bumjoon Seo, Won Bo Lee

30b Uniformly Embedded Cobalt Phosphide Nanoparticles in Carbon Nanofibers As Sulfur Host Via Co-PBA in-Situ Growth for High-Performance Li-S Batteries 79
Zhi Liu, Cameron Romero, Kenneth Gordon, Le'Naedrea Moore, Ling Fei

30d High Resolution Measurement of Potential-Dependent Electrochemical Activities on HOPG Using Scanning Electrochemical Cell Microscopy (SECCM) 80
Myunghoon Choi, Sang-Joon Cho, Lane Baker, Stefan Kaemmer

30e Wetting Transparency of Free-Suspended Single-Layer Graphene on Liquid Substrate. 81
Fan Yang, Annette Thompson, Lei Li

30f Feedback Control System of Iron Catalyst in Reactors Using Deep-Injection Floating Catalyst Chemical Vapor Deposition (DI-FCCVD) 82
Mingrui Gong, Anavi Benavides Figueroa, Arthur Sloan, Matteo Pasquali

30g Facile Synthesis of Halogen Terminated Mxene from Different MAX Phases for Electromagnetic Interface Shielding 83
Shane Reed, Pitchaimari Gnanasekar, Sanjay Singh, Natalie Arnett, Dinadayalane Tandabany, Theo Siegrist, Subramanian Ramakrishnan

30h Rheo-Electric Investigation of Carbon Black Suspensions Undergoing Shear-Induced Microstructural Rearrangement 85
Paolo Ramos, Julie Hipp, Jeffrey Richards

30i Lyotropic Liquid Crystals of Cellulose Nanocrystal/Boron Nitride Nanotube Mixtures: Phase Behavior and Aligned Assemblies 86
Eren Katnas, Geyou Ao

30j Scalable Manufacturing of Inorganic Nanomaterials Using Jet Mixing Reactors 87
Priya Jana, Faiz Khan, Pinaki Ranadive, Jessica Winter, Nicholas Brunelli

30k Threshold and Excess Electrolyte-to-Sulfur (E/S) Ratios in Lean-Electrolyte Li-S Batteries 88
Xiaosi Gao, Siddharth Tiwari, Aditya Pandey, Yiqi Shao, Ziang Gao, Yong Lak Joo

30l Glycopolymer-Wrapped Carbon Nanotubes Detection of Viral Proteins 89
Brandon Heppe, Geyou Ao, Xue-Long Sun

30n Polymer-Cerium Oxide Nanocomposites Synthesized By Hydrothermal Method for Bone Tissue Regeneration Applications 90
Shreya Pawar

30q Lipid Nanoparticle Structure and Immune Response Regulate mRNA Organ Tropism and Pup Growth during Pregnancy in Mice. 92
Namit Chaudhary, Alexandra Newby, Mariah Arral, Saigopalakrishna Yerneni, Samuel Lopresti, Rose Doerfler, Daria Petersen, Bethany Fox, Tiffany Coon, Angela Malaney, Yoel Sadovsky, Kathryn Whitehead

30r The Particle Drifting Effect – Impact of Membrane, Sink Condition, Particle Size, Drug Solubility, and Colloid Type 93
Na Li

POSTER SESSION: NSEF GRADUATE STUDENT POSTER COMPETITION

118a Biodegradable Polymer Capsule for Long-Term Drug Release 94
Zhiwei Zheng, Thomas Waterkotte, Tilahun Debele, Gregory Dion, Yoonjee Park

118b Magnetic Nanocomposites for Environmental Remediation 95
Pranto Paul, Marissa Nicholson, J. Zach Hilt

118c Upcycling of Waste Polyvinyl Chloride (PVC)-Based Materials to Electrospun Nanofibers for Water Treatment 96
Atta Ur Razzaq, Milad Esfahani

118d Design and Properties of Magnetic Particle Imaging (MPI) Tracers Using Nitrodopa Anchored Coatings. 97
Neel Eswaran, Hayden Good, Ambar Velazquez Albino, Eric Imhoff, Carlos Rinaldi-Ramos

118e Automated Fe Quantification Assay to Accelerate Magnetic Nanoparticle Studies 98
Eric Imhoff, Ambar Velazquez Albino, Esteban Bermudez-Berrios, Carlos Rinaldi-Ramos

118f Evaluating the Quality Control of Aqueous Carbon Nanotube Dispersions Using UV-Vis Spectroscopy and Zetasizer Analysis 100
Victoria Stege, Chelsea Monty-Bromer, Shelby Daniels, Michael Fulmer, Zachary Cheney

118g Surfactant Mediated Phase Transfer of Iron Oxide Nanoparticles 101
Esteban Bermudez-Berrios, Ambar Velazquez Albino, Eric Imhoff, Carlos Rinaldi-Ramos

118h Corona Phase Molecular Recognition of the Interleukin-6 (IL-6) Family of Cytokines Using Nir Fluorescent Single Walled Carbon Nanotubes 102
Xiaojia Jin, Xun Gong, Volodymyr Koman, Daniel James Lundberg, Michael Strano

118i Rational Synthesis, Morphology, and Propertyrelationship of Non-Precious Bi-Metallic Hybridnanocomposites for Improving Oxygen Reductionreactions 103
Mahshid Mokhtarnejad, Mariana Milano-Benitez, Bamin Khomami

NANOSCALE SCIENCE ENGINEERING FORUM (NSEF): FACULTY CANDIDATE

319a Using Liquid Crystals to Control Molecular Packing in Organic Solids over Macroscopic Dimensions 104
Kushal Bagchi, Paul F. Nealey

319b Exploring the Potential of Smart Supramolecular Materials for Energy and Beyond 105
Shuhao Liu, Yu-Ting Lin, Bhargavi Bhat, Mustafa Akbulut

319c Holistic Understanding of Nanocrystal Stability 106
Dr. Aniket Sandip Mule

319e Macroscopic Evidence of the Liquidlike Nature of Nanoscale Polydimethylsiloxane Brushes 108
Xiaoxiao Zhao

NANOSCALE SCIENCE ENGINEERING FORUM (NSEF)

262b Liquid Crystalline Phase Behavior of Boron Nitride Nanotube Aqueous Dispersion 109
Joe Khoury, Eldar Khabushev, Lyndsey R. Scammell, Cheol Park, Angel Marti-Arbona, Matteo Pasquali

262c Rational Synthesis, Morphology, and Property Relationship of Non-Precious Bi-Metallic Hybrid Nanocomposites for Improving Oxygen Reduction Reactions 110
Mahshid Mokhtarnejad, Dibyendu Mukherjee, Bamin Khomami

262d Elucidating the Role of Surface Functionalization on the Photoluminescent Properties of Group IV Nanocrystals 111
Maharram Jabrayilov, Matthew Panthani

262e Deep Learning-Based Segmentation of Complex Microparticles in Scanning Electron Microscopy Images 112
Anastasiia Visheratina, Alexander Visheratin, Prashant Kumar, Michael Veksler, Nicholas Kotov

262f Minimizing the Formation of Polynuclear Palladium Hydroxo Complexes in the Mineralization of Barley Stripe Mosaic Virus 114
Che-Yu Chou, Akash J. Vaidya, Chamath Vindula Basnayake Pussepitiyalage, Shohreh Hemmati, Kevin Solomon, Sue Loesch-Fries, Michael T. Harris

262g Atomically Precise Nanoclusters with Reversible Isomeric Transformation for Rotary Nanomotors 115
Zhaoxian Qin, Gao Li

262h Functionalization of Silicon with Tunable Bilayered Molecular System for Subsequent Photopatterning and Phosphorus-Based Dopant Attachment 116
Patrick R. Raffaelle, Alexander Shestopalov

NANOSTRUCTURED BIOMIMETIC AND BIOHYBRID MATERIALS AND DEVICES

426a DNA Patterning on Carbon Nanotube-Based Sensors Resolved By X-Ray Scattering Interferometry 117
Rebecca L. Pinals, Daniel J. Rosenberg, Francis J. Cunningham, Joshua Hubbard, Natalie S. Goh, Jeffrey Wang, Emily Hayman, Gregory Hura, Markita Landry

426b Programming DNA Self-Assembly for Molecular Robotics and Responsive Materials 118
Alexander Marras

426c Enhancing Clot Formation and Mechanical Strength with Graphene / Peg Hemostatic Nanoplates 119
Tannaz Farrokhi, Manos Gkikas

426d Nanostructured, Biocompatible Mxene-Based Hybrid Membranes Fabricated Using Vacuum-Assisted Layer-By-Layer Assembly 120
Gelareh Rezvan, Farivash Gholamirad, Mary Walden, Monirosadat Sadati, Nader Taheri-Qazvini

30t Next generation nano-diagnostic devices: Graphene interaction with skin cancer cells 121
Kartikey Sharma

426f Scalable Nanofabrication of Biomimetic Broadband Antireflection Coatings 122
Fan Gao, Jiachen Wu, Peng Jiang

426g Protein Aggregation on Metal-Oxide Nanozymes Governs Activity and Cellular Uptake 123
Robert Nißler, Lena Dennebouy, Alexander Gogos, Michael A. Pais, Anna L. Neuer, Monika Zimmermann, Ioana Lese, Inge Herrmann

426h Direct Observation of Heterogeneous Distribution of Enzymes in Metal-Organic Frameworks and Its Effects in Propelling Nanomotors 124
Yu Liu, Jun Ge

2D AND 3D NANOMATERIALS

158a One-Step Fluorine-Free Synthesis of Delaminated, OH-Terminated Ti_3C_2: High Photocatalytic NO_x Storage Selectivity Enabled By Coupling TiO_2 and Ti_3C_2-OH 125
Ahmed Al Mayyahi, Swagotom Sarker, Brian M. Everhart, Xiaoqing He, Placidus Amama

158b Controlled Synthesis of 2D Transition Metal Dichalcogenides for Electronic Nanosensors 126
Meng-Qiang Zhao

158d Chemical Synthesis and Physical Properties of Graphimine - a New 2D High-Performance Polymer 127
Kätchen K. Lachmayr, David C. McLeod, Eric D. Wetzel, Steven Lustig

158e The Effects of Ionic Strength on Rheological Properties and Microstructural Changes in Aqueous Mxene ($Ti_3C_2T_x$) Dispersions 128
Mackenzie Woods, Fatima Hamade, Majid Beidaghi, Virginia Davis

158f Programming Self-Assembly of Plasmonic Nanomaterials with Sequence-Controlled DNA Polymers 129
Yifeng Shi, Grigory Tikhomirov

BIONANOTECHNOLOGY FOR SENSORS AND IMAGING II

352a Tracking Dendritic Cell Vaccination Using Magnetic Particle Imaging 130
Bo Yu, Ginger Moore, Alexander Armstrong, N'Quel Bonner, Sitong Liu, Ambar Velazquez Albino, Duane A. Mitchell, Carlos Rinaldi-Ramos

352b Facile and Quantitative Detection of Nucleic Acid Markers Via Conformational Change of Gold Nanoparticle Assemblies 132
Yuto Sahara, Noriko Nakamura, Seiichi Ohta

352c Detection and Visualization of HER2 Clustering Using an Aggregation Induced Emission Based Fluorescent Probe Molecule 135
Yasaman Moradi, Jerry S. H. Lee, Andrea M. Armani

352d DNA-Mediated, on-Membrane Sequential Assemblyof Conjugated Polymer Nanoparticles for Sensitive Detectionof Cell Surface Markers 137
Yuki Maeda, Noriko Nakamura, Seiichi Ohta

352e Improve Sensitivity and Selectivity of Neurotransmitter Sensing through UV Plasmonics Enhanced Native Fluorescence 140
Mohammad Mohammadi, Yunshan Wang

352f Field Flow Fractionation-Based Analysis for Evaluating Slight Aggregation of Surface-Functionalized Nanoparticles That Can Impact Their Biological Interaction 141
Hiroki Tsuchiya, Noriko Nakamura, Seiichi Ohta

352g An Injectable Hyaluronic Acid-Based Nano-Contrast Agent for CT Detection of Inflammation and Osteoarthritis 144
Ioanna Chatzaki, Alice Truong, Mary Ruskowski, Manos Gkikas

BIONANOTECHNOLOGY FOR GENE AND DRUG DELIVERY II

229a Mucus Penetration of Surface-Engineered Nanoparticles in Various pH Microenvironments. 145
Yiyang Guo, Bingbing Sun

229b Modulating Nonspecific Uptake of Engineered Extracellular Vesicles 147
Beth Dibiase, Roxana E. Mitrut, Taylor Gunnels, Neha Kamat, Joshua Leonard

229c Bubble-Based Microrobots for Epithelial Pinning and Sustained Drug Delivery 149
Jin Gyun Lee, Ritu Raj, Cooper P. Thome, Nicole Day, Payton Martinez, Nick Bottenus, Ankur Gupta, C. Wyatt Shields Iv

229d On-Demand Drug Release of Polymeric Core-Shell Nanoparticles Developed through Aqueous Surface-Initiated Sara-ATRP 151
Nahieli Preciado, Colin Stark, Jingyi Liu, Todd Hoare

229e Supramolecular Peptide-Protein Granules for Intracellular Protein Delivery 154
Renjie Liu, Madisen Domayer, Alex Adolphson, Greg Hudalla

229f Desirable Formulations for Inhalable Particulate Immunotherapeutics Using Model Peg-Based Nanoparticle 155
Emma Sudduth, Catherine Fromen

229g Macrophage Internalization of PEGDA Nanoparticles Drives Secretion of Pro-Survival Signals 156
Michael Trautmann-Rodriguez, Simone Sabnis, Catherine Fromen

229h pH-Mediated Mucus Penetration of Zwitterionic Polydopamine-Modified Silica Nanoparticles 157
Yubin Ma, Bingbing Sun

Author Index

3a Micro/Nanoengineered Adhesive Biomaterials for Regenerative Medicine and Drug Delivery

Nasim Annabi

Tissue engineering (TE) is a rapidly expanding field combining knowledge in biological sciences, engineering, and medicine. The main goal of TE is to develop innovative strategies for engineering tissue substitutes *in vitro, ex vivo,* and *in vivo* to restore, maintain, or promote normal tissue function. Biomaterials have played an important role in the development of tissue constructs that mimic the structures, and physical and biological properties of the native tissues. However, there are still many challenges in micro and nano-engineering of biomimetic biomaterials for the development of fully functional tissue constructs. The combination of advanced biomaterials, with tunable properties, with micro- and nanoscale technologies have been shown to hold great potential to address the current challenges in tissue engineering. Our research focuses on the design and synthesis of elastic biomaterials using recombinant proteins and chemically modified naturally derived biopolymers. The engineered elastin-based biomaterials are merged with different types of nanomaterials and micro/nano-fabrication technologies to form multi-functional hydrogels for a wide range of applications in tissue engineering and drug delivery. In addition, we have developed new chemistries to improve the wet tissue adhesion and toughness of these hydrogels and use them as bioadhesives and sealants for sealing and repair of soft and elastic tissues such lung, heart, skin, and cornea. Our engineered tough bioadhesives can overcome many challenges in the development and practical translation of hydrogel-based biomaterials including robust, reliable, and functional integration to wet, dynamic, and living biological tissues. Integration of innovative nano-based drug delivery systems with these bioadhesives has enabled us to expand their applications for treatment of various ocular diseases via targeted and controlled delivery of therapeutics. In this presentation, I will outline our recent work on the design and microengineering of bioadhesive hydrogels for soft tissue sealing and regeneration along with their clinical applications as matrices for drug delivery and targeted therapy.

3b Harnessing Chaos: Rational Design of Nanoscale Surfaces for Cavitation-Based Imaging Agents, Nanomotors, and Protein Therapeutics.

Andrew Goodwin

Cavitation events, or the formation of a gas pocket in a liquid medium, can be highly destructive and disruptive. These events offer a unique opportunity to localize large amounts of energy using remote triggering. We have sought to understand how nanoscale surfaces can facilitate or inhibit cavitation by examining the structure-property relationships that give rise to efficient bubble nucleation. These approaches in turn underlie our new technologies in imaging, therapy, and protective coatings.

First, the formation of bubbles in solution can be used as efficient ultrasound contrast agents. Because of their high compressibility and low density, bubbles offer excellent contrast in aqueous media. However, microbubbles suffer from low stability and an inability to leave the bloodstream. In this work, we engineered functionalized silica nanoparticles that could facilitate the formation of transient microbubbles in response to an incoming acoustic wave.

Second, acoustic cavitation on nanoparticle surfaces gives rise to many therapeutic applications. Non-equilibrium bubble collapse is a violent process that concentrates both thermal and mechanical energy capable of not only killing cells but transporting nanoparticles through dense media such as extracellular matrix. Notably, the propulsion of nanoparticles can kill cells up to hundreds of microns away, thereby addressing transport limitations in solid tumors.

Third, we sought to rationally design surfaces that could retard cavitation as well. Solutions of protein therapeutics are known to form particulates upon impact and resulting cavitation. These particulates can in turn induce an anaphylactic response in patients, in some cases leading to patient death. Here, new coatings were applied to standard glass vials to reduce their tendency towards cavitation, thereby protecting the proteins in solution from irreversible aggregation.

3c Building Bioactivity and Nanostructure into Slippery Liquid-Infused Porous Surfaces

David M. Lynn

Slippery liquid-infused porous surfaces (or 'SLIPS') and other types of liquid-infused materials provide new approaches to prevent biofouling on commercial and industrial surfaces, including those used to design biomedical devices. Many different types of SLIPS fabricated using porous polymer matrices and hydrophobic oils have been demonstrated to resist adhesion and colonization by microorganisms. However, while SLIPS are generally good at preventing fouling on surfaces to which they are applied, they can generally do little to prevent proliferation of non-adherent organisms —*e.g.*, to stop them from colonizing other surfaces or prevent them from engaging in other behaviors, such as virulence factor production, that could also lead to infection or fouling.

We have developed multi-functional bioactive SLIPS that address these issues and expand the potential utility of slippery surfaces in antimicrobial contexts. Our approach is based on the incorporation and controlled release of small-molecule antimicrobial agents from hydrophobic porous polymer matrices used to host infused oil phases. Our studies reveal that SLIPS designed using nanoporous coatings fabricated by covalent layer-by-layer assembly or other polymer spinning-based methods can prevent short- and longer-term colonization and biofilm formation by common fungal and bacterial pathogens. We demonstrate that both the polymer and liquid phases comprising these materials can be exploited to load and sustain the release of antimicrobial agents and other types of small- and large-molecule actives into surrounding media. Recent strategies to exploit the properties of nanostructured liquids, such as water-in-oil nanoemulsions, as infused phases and design antifouling SLIPS capable of hosting and releasing highly water-soluble agents will also be discussed. These approaches improve the inherent anti-fouling properties of these materials, enable them to more efficiently kill planktonic pathogens, and can endow them with other useful functional properties.

These approaches are modular and can be used to fabricate multi-functional, bioactive SLIPS on complex surfaces, including the luminal spaces of flexible polymer tubing of the type used to fabricate catheters. This strategy thus has the potential to be general and can be exploited to release a broad range of oil- and water-soluble cargo. We anticipate that these materials, strategies, and concepts will enable new designs of bioactive slippery surfaces with improved anti-fouling properties and open the door to new applications of slippery liquid-infused materials that host or promote the release of other types of bioactive agents. Efforts to apply these approaches to the design of anti-biofouling tubing and coatings that can identify and report on virulence phenotypes or attenuate communication in non-adherent bacteria will also be discussed.

37a Exploring the Best CVD Conditions for Growth of Small-Diameter Single-Wall Carbon Nanotubes Using an Autonomous Research System

Brian M. Everhart, Rahul Rao, Tsung Wei Liu, Diego Gomez Gualdron, Benji Maruyama, Placidus Amama

Due to the breadth of parameters that affect growth of single-wall carbon nanotubes (SWCNTs) via chemical vapor deposition, rapid experimentation is a powerful tool for investigating conditions for selective growth of small-diameter SWCNTs. Here we utilize an Autonomous Research System (ARES)—an automated, high throughput, laser-induced CVD system with in situ Raman spectral feedback—to study the roles of Ru promotion of Co catalyst and type of feedstock in the growth of small-diameter SWCNTs at different temperatures. We demonstrate through over 200 growth experiments in ARES and ex-situ multi-excitation Raman spectroscopic characterization that Ru-promoted Co catalyst nearly doubles the selectivity of small-diameter SWCNTs (diameters below 1nm) between growth temperatures of 650°C and700°C in comparison to pure Co. At elevated temperatures between 800 and 850°C, Ru stabilizes Co catalyst nanoparticles and increases the selectivity of small-diameter SWCNTs by almost a factor of three. Results reveal SWCNT diameters are not only dependent on the size of the catalyst but also on the precursor chemistry as selectivity towards small-diameter SWCNTs decreases in the following order: ethylene > acetylene (1% in He) > FTS-GP (Fischer-Tropsch synthesis gaseous product mixture). Density functional theory (DFT) simulations using 13- and 55-atom clusters with similar Ru/Co ratios provide mechanistic insight into the observed phenomenon, revealing increased cohesive energies of Co clusters after the addition of Ru irrespective of the location. Our findings indicate the presence of ~10% Ru in Co increases the sintering resistance and stability of small nanoparticles, as well as selectivity towards small-diameter SWCNTs.

37b Molecular Design and Investigation of 2-D Hybrid Materials

Sanket Deshmukh

2-Dimensional (2-D) materials and their composite hybrids have swiftly risen as pivotal materials, finding relevance in diverse sectors, from tribological to biomedical arenas. However, tailoring these hybrid 2-D materials to achieve specific structures, properties, and functionalities remains a formidable task. In this talk, I will delve into our recent work in the computational design of layered hybrid materials, focusing on graphene, hexagonal boron nitride, and molybdenum disulfide. Utilizing all-atom molecular dynamics (MD) simulations, we demonstrate that strategic alterations in the sequencing of these layered materials can modulate their hydration levels. Subsequently, I will shed light on the intricate molecular interactions that underpin the formation of graphene nanoscrolls when intertwined with thermosensitive polymers. Through our coarse-grained molecular dynamics (MD) simulations, we've identified that a delicate interplay of non-bonded interactions between the polymer and graphene is crucial for the formation of stable nanoscrolls. Collectively, our findings offer a promising roadmap for the conceptualization and creation of innovative hybrid materials.

370a Award Submission: Tracking Dendritic Cell Vaccination Using Magnetic Particle Imaging

Bo Yu

Dendritic cell (DC) based vaccines have proven to be a safe therapeutic approach, although with inconsistent clinical results. DCs must migrate to lymphoid organs to be effective and the magnitude of anti-tumor response is expected to be correlated to DC migration to lymph nodes (LNs). It's reported that a small fraction of administered DCs migrate to LNs. Therefore, non-invasive methods to sensitively monitor and quantify migration of adoptive DC therapies would be tremendously helpful in evaluating immune response. Although techniques such as magnetic resonance imaging (MRI), single-photon emission computed tomography (SPECT), and positron emission tomography (PET) have previously been used to track DCs, these techniques are limited by low cell detection sensitivity in the range of 10^4 cells or more.

Magnetic particle imaging (MPI) is a novel molecular imaging technology that can sensitively and quantitatively detect iron oxide tracers *in vivo*, with higher cell detection sensitivity compared to other imaging approaches. Cell detection sensitivity in MPI is determined by two factors: the properties of the tracer used to label cells and the extent of tracer uptake per cell. Other than detection sensitivity, to be able to differentiate signals from LNs and injection sites, resolution and imaging conditions are crucial factors in tracking DC migration *in vivo*.

In this study, cellular uptake was evaluated for MPI tailored tracers (RL-1) coated with poly(maleic anhydride-alt-1-octadecene) (PMAO). The MPI maximum signal intensity for the tracer was 101 $mgFe^{-1}$. The upper labeling limit for marrow-derived dendritic cells (BMDCs) was determined to be ~ 50 pgFe/cell by screening SPION concentration and incubation time. Cell viability, functionality, and surface biomarker expression were not affected by tracer labeling. Confocal images suggested particles are in vesicle-like structures inside cells. *In vitro* tracer fate studies suggested particles are not degraded or exocytosed after 3 days of culture. Cell dilution studies demonstrated detection of as little as 10^3 cells with MPI using optimized labeling conditions. An *in vivo* BMDC tracking study was performed using high sensitivity and high-sensitivity/high-resolution scan modes and different administration routes (foot pad subcutaneous injection and hind flank intradermal injection). Administered DC numbers were quantifiable using *in vivo* day 0 MPI measurements. BMDC migration quantified by *ex vivo* lymph node MPI measurements ranged from 3000 to 5000 cells, however, signals in LNs *in vivo* were not detectable because the strong signal from the injection site masked the weak LN signal. An *ex vivo* phantom study was performed to evaluate signal differentiation and quantification accuracy as a function of separation distances, administered doses, and imaging settings. Then an *in vivo*BMDC tracking experiment using advanced imaging settings with partial filed of view and full field of view was performed, in vivo BMDC migration was trackable with partial field of view imaging. The correlation of in vivo LN MPI signal and ex vivo LN MPI signal was evaluated. BMDC migration efficiency was compared with and without GM-CSF. A survival study will be performed to evaluate correlation of LN MPI signal and DC vaccination therapeutic response. Together, these studies demonstrate the potential of MPI as tool for quantitative tracking of DC migration in DC vaccination studies.

370b Award Submission: Accelerated Blood Clearance of Lipid Nanoparticles Occurs Regardless of Formulation

Mariah Arral

Lipid nanoparticle (LNP) based mRNA delivery has the potential to revolutionize medicine as it can be used in vaccines, protein replacement, and cancer therapy. Most viable mRNA therapies will need repeated administration to enact a continuous therapeutic effect. Unfortunately, some LNP formulations lose mRNA delivery efficacy upon repeat dosing due to Accelerated Blood Clearance (ABC). ABC is a tachyphylaxis response against the PEGylated lipid component of LNPs, and correlates with anti-PEG antibody levels in serum. The goal of this study is to investigate whether formulation optimization of an ABC-inducing LNP can enable successful repeated dosage. To this extent, we investigated the effect of different PEGylated lipids, PEG molar ratios, and administration routes on repeat efficacy of an ABC-inducing LNP and the adaptive immune response observed after delivery. Lipid nanoparticles (LNP A) encapsulating Erythropoietin (EPO) mRNA were formulated with varied PEGylated lipids and molar ratios (0-2.5%). Mice were injected intravenously, intramuscularly, or intraperitoneally at a dose of 0.5 mg/kg, two weeks apart. EPO protein expression and serum antibody titers were measured using ELISA. ABC-inducing LNP A elicited efficacy loss after repeat dose regardless of administration route and prompted anti-PEG IgM production that peaked 21 days post injection. Next, we explored if changing the PEG-lipid type would allow for LNP A to successfully repeat delivery. Of the three PEG-lipids tested, none resulted in successful repeated dosage, and all significantly increased anti-PEG IgM on day 21 compared to baseline. Following these findings, we then changed the concentration of PEG-lipid used. Loss of efficacy was seen for LNPs with 1.0-2.5 mol% PEG-lipid. However, LNPs with less than 1.0 mol% PEG-lipid loss their ability to efficaciously deliver EPO. Formulation that showed significant EPO delivery over PBS did have increased anti-PEG IgM on day 21. Our work indicates that altering the PEG-lipid type, PEG-lipid concentration, or route of administration does not alleviate an ABC-inducing LNP's efficacy loss. Loss of efficacy correlated with elevated anti-PEG IgM levels, suggesting that ABC of LNPs is associated with anti-PEG IgM. However, the reason why anti-PEG IgM increases and causes ABC occurs with one LNP and not another remains unknown. Future work will evaluate the effect of pre-existing anti-PEG levels on *de novo* efficacy of an FDA-approved LNP. It is crucial to determine if ABC results in elimination of any LNP to successfully deliver mRNA, as it could have clinical implications for future therapeutics.

370c Award Submission: Mucus Penetration of Surface-Engineered Nanoparticles in Various pH Microenvironments.

Yiyang Guo

The inhaled medications delivered by nano- or micro-sized carriers are widely used for the prevention and treatment of lung diseases such as cystic fibrosis (CF), chronic obstructive pulmonary disease (COPD), asthma, and COVID-19. During the transport of the drug carriers in the airway, the mucus lines the surface of the airway and hinders the effective delivery of active ingredients. The mucus layer is composed of 98% water, 1.6% mucin, and 0.4% of the remainder, *e.g*., cellular debris, salts, DNA, lipids, globular proteins, *etc*. Mucin form a three-dimensional network to restrain the movement of particles *via* the size exclusion and interaction filtration effect. It has been demonstrated that proper control of physicochemical properties of carriers could improve the transport and fate of medications.

The neutral charge and hydrophilic surface have been identified as critical requirements for the effective trans-mucus drug carriers. Poly (ethylene glycol) (PEG) has been used as a stealth polymer that repels protein adsorption to assist the drug carriers to escape the binding of mucin. Their molecular weight, chain length, and content have been shown to affect mucus penetration performance. The zwitterionic design mimics the surface characteristics of the virus surface, and further exhibits the equal densities of positive and negative charges without the presence of hydrophobic patches. The surface modification by zwitterionic polymer and density adjustment enables enhanced mucus penetration of particles. Furthermore, the particle surface modification by block copolymer (F-127), peptides and protein were potentially alternative strategies to improve penetration ability of particles. Although PEGylation is often considered as a "gold standard" to assess mucus penetration, and various surface modification strategies have been proposed. the mechanistic understandings of particle surface property-mediated mucus penetration are currently missing.

The mucus microenvironment changes according to the physiological and pathological states. In asthma exacerbation, healthy, and airway infection or inflammation states, the pH values of mucus is altered. However, the airway mucus penetration model in current study is mainly derived from commercial porcine gastric mucin. Regardless of the mucin suspension or the artificial mucus, the pH value of the system is acidic (pH=3-5), which is different from the natural airway health neutral mucus environment. Although sputum taken from patients such as cystic fibrosis (CF) as a surrogate is similar to the mucus environment experienced by drug carriers after airway administration, factors such as the causes (endogenous or exogenous exposure) and states of diseases (mild or severe) were not considered. Acid respiration occurs in exacerbation of CF (mucus pH=5.3) and asthma (pH=5.2), but the pH of mucus is increased in mild asthma (pH=7.6) and stable CF (pH=5.9). Moreover, the mucus is an alkaline environment (pH=7.8-8.5) in inflammatory diseases. Therefore, mucus pH microenvironment should be taken into consideration during inhalation delivery.

In this study, we prepared a comprehensive library of engineered silica nanoparticles (SNPs) with controlled surface ligand type and density. According to the size exclusion mechanism of mucus, nanoparticles with a size range of 200-500 nm in diameter are easier to penetrate the mucus barrier. Therefore, the 300 nm SNPs were selected to study their mucus penetration ability. The pristine SNPs were synthesized using the Stöber method, and SNPs were doped with fluorescein isothiocyanate (FITC) for particle tracking. Then, SNPs were further modified with silane coupling agents to prepare a library of engineered SNPs with controlled functional groups and ligand densities. Then, particles were dispersed in the airway mucus with mimetic physiological and pathological pH values to evaluate particles movement behavior by using multiple-particle tracking (MPT). PEGylated and amine SNPs showed pH-independent mucus penetration and entrapment, the carboxyl-SNPs exhibited movement enhancement only in weakly alkaline environments. The mechanisms were further systematically analyzed from three aspects, *i.e.*, mucus properties, SNPs surface properties, and mucus-SNPs interactions, respectively.

A recent study has revealed that mucin undergoes a sol-gel transition in acidic-neutral environments, and the cross-linking properties of mucus are stronger in acidic than that in neutral condition. In our study, it was demonstrated that the pH-dependent changes in mucus macro-rheology and micro-structure were the main reasons for the observed variations of particle motion behavior. From the perspective of mucin micro-structure, the changes in the macro-crosslink properties of mucus at different pH was due to the changes in the microstructure of mucin. During the acidification of mucus, carboxylates on mucin were protonated. As a result, the electrostatic interaction between carboxylate and amine groups were compromised, and hydrophobic domains hidden in salt bridge folded structure were exposed, promoting the cross-linking of mucus gel. In addition to the influence of the hydrophobic domain, electrostatic interactions due to the exposure of mucin glycosyl side chains also play an essential role. In generally, mucin is negatively charge due to the sialic and sulfate components in the glycosyl side chain. In the present study, this pH-dependent conformation and charge transition were observed. The circular dichroism (CD) analysis revealed that mucin was dominated with the random coil conformation, and the pH-dependent intensity and width changes at 206 nm was observed, suggesting the conformation undergoes a transition from isotropic random coil to anisotropic extended random coil with the decrease of pH values. Additionally, the mucin fiber in acidic environment exhibited increased zeta potential, thus enhancing electrostatic interactions between mucin fibers, increasing the viscosity of the mucus and inhibiting the movement of the particles.

Furthermore, isothermal titration calorimetry (ITC) reveals that the independent of the functional groups and mucus pH, the interaction of mucin with SNPs were exothermic and entropy loss process, indicating an enthalpy-driven interaction and the formation of non-covalent bonds was reason why particles were trapped in mucus, while the dense brush structure formed by PEG modification shielded the interaction between particles and mucin to achieve mucus penetration. Furthermore, to study the effect of mucus pH, the amine-modified SNPs were selected to titrate with mucin solutions. The association constant (K_a) of SNPs in acidic (pH 5.2) environment was an order of magnitude higher than those in neutral (pH 7.0) and slightly alkaline (pH 7.7) ones. The stronger SNP-mucin interaction was correlated with the fact that particles being trapped in the in acidic mucus. Therefore, this study demonstrated that the fate of drug carrier particles in airway mucus was pH microenvironment, particle surface ligand type, and density-dependent, and provides an engineered approach to control the mucus penetrating capabilities of drug delivery carrier for the personalized treatment and prevention of pulmonary diseases.

370d Award Submission: On-Demand Drug Release of Polymeric Core-Shell Nanoparticles Developed through Aqueous Surface-Initiated Sara-ATRP

Nahieli Preciado

Chemotherapeutic delivery vehicles that can be externally addressed to release drugs 'on-demand' offer significant potential to dynamically adjust treatment according to the patient's response (i.e. personalized medicine) without requiring repeated invasive and often painful drug administrations. Even though new drug delivery methods that lower secondary risks have been developed via nanomedicine, they have not significantly improved patients' survival rates [1]. New therapeutic paradigms are required to achieve the desired improvements in patient outcomes.

One key group of materials with potential in this regard is nanoscale stimuli-responsive systems that react to either exogenous stimuli such as temperature variations, magnetic fields, ultrasound, light, or electric fields [2] or endogenous changes in local pH and enzyme concentration associated with a disease [3]. Magnetic triggering is especially attractive as an external trigger since magnetic fields are highly penetrative (>1 m in all tissues [4], as opposed to light and, in specific scenarios, ultrasound, that have penetration depths as short as the millimetre scale) without adversely affecting biological tissues at biologically-relevant field strengths/frequencies. The most common implementation of magnetic triggering applies superparamagnetic iron oxide nanoparticles (SPIONs), whose single-crystal nanostructure enables the transformation of energy from an alternating magnetic field (AMF) into heat through Neel and Brownian relaxation mechanisms. SPIONs can be used alone for cancer therapy by leveraging this heat generation capacity to induce localized hyperthermia to kill cancer cells, with thermotherapy noted to have fewer side effects than chemo or radiotherapy [5]. However, increasing recent interest has focused on creating hybrid nanosystems that merge inorganic nanoparticles with drug delivery systems to **synergistically** induce apoptosis in cancer cells, with the drug acting together with the higher local temperature to create a more efficient cancer therapy [6], [7].

Introducing the potential for *pulsatile*, rather than simple diffusion-controlled, release would add another therapeutic option to trigger release of drug "on-demand" as per a patient's need, addressing the lack of control over release kinetics currently achievable once a drug delivery vehicle is administered into the body. Thermoresponsive hydrogels have been previously used for this purpose; however, their high porosity allows loaded drug to be released rapidly coupled with relatively high "off" state drug release by diffusion through even the collapsed gel phase. As an alternative, the use of the glass transition (Tg) properties of solid polymer-based particles would significantly reduce "off" state release while still allowing for temperature-induced "on" state release enhancements if the Tg is tuned to be just above physiological temperature. While current research has successfully led to the development of a range of injectable smart materials on many length scales and with many geometries, the capacity to safely and non-invasively control the smart response(s) locally and repeatedly without biological side-effects over the long term, remains challenging. By combining the glass transition with core-shell nanoparticles that can enable stabilization/targeting of the nanotherapeutic, we see the potential to develop an injectable drug delivery vehicle with the capacity for repeated pulsing while avoiding substantial drug delivery in the 'off' state that currently hinders the translation of on-demand drug delivery devices.

Objective

We aim to develop a nanoscale injectable on-demand drug delivery system that can be activated by heat and/or magnetic fields to selectively deliver drug only when triggered. Such a system will decrease the secondary effects of systemic cancer therapy drugs, increasing drug concentration at the target site, and reduce total drug dose compared to the systemic delivery approach. The gel shell around the glass transition-switchable nanoparticle promotes colloidal stability and reduces biological responses. The heat-induced trigger will allow for on-demand release of only the needed amount of drug only when heat or an alternating magnetic field is applied, the latter enabled by the incorporation of SPIONs in the shell of the particles.

Methodology

The core nanoparticles were fabricated by free radical polymerization of methyl methacrylate (MMA), butyl methacrylate (BMA), and 2-(2bromoisobutyryloxy)ethyl methacrylate BIEM) using an oil-in-water miniemulsion technique, using rhodamine B as the model drug loaded by directly incorporating it during the miniemulsion process. The particle size was determined via dynamic light scattering. Following, the bromine groups in BIEM were used as an initiator to grow the OEGMA shell via aqueous surface-initiated SARA-ATRP. SPION groups were incorporated in the shell. Zwitterionic polymers functionalized with either ketone or hydrazide polymers (enabling the formation of a hydrazone bond upon contact) were used to form an *in situ*-gelling hydrogel to immobilize the core-shell particles at a specific injection site, with the polymer concentration adjusted to obtain a desirable gelation time (1-2 min). The core-shell particles were dispersed in one or both precursor polymer solutions and then co-injected through a double barrel syringe fitted with a static mixer to form the nanocomposite hydrogels for drug release testing. Rhodamine B release was assessed by immersing the nanocomposite gels in PBS and exposing them to heat (45°C) or a 200 kHz alternating magnetic field for activation.

Results

Core nanoparticles were formulated using different ratios of monomers and macroinitiators to determine the effect the macroinitiator would have on the final glass transition temperature. Increasing the concentration of the macroinitiator from 0.25% to 1.4% only decreased the Tg by 2°C (58.5°C to 55.7°C), too high for safe triggering under physiological conditions. As such, to obtain a desirable Tg (45°C). butyl methacrylate was added to the MMA-BIEM mix, decreasing the Tg by 10°C and obtaining a final Tg of 45.6°C. The particle sizes were not significantly affected by the comonomer composition, yielding core particles of ~120 nm in diameter across a range of monomer contents.

The SARA-ATRP reaction was subsequently optimized to grow the passivating polymeric shell around the core nanoparticle. Different parameters such as the polymerization temperature and the amount of macroinitiator/BIEM were changed to determine the best conditions for shell growth. Using sub-ambient temperatures or higher concentrations of macroinitiator promoted interactions between the particles, creating a particle network rather than individual core-shell nanoparticles. In contrast, performing the reaction at room temperature and decreasing macroinitiator concentration resulted in stable core-shell nanoparticles. Shell growth was analyzed using dynamic light scattering after different polymerization times to control the shell thickness, allowing for the creation of core-shell nanoparticles of ~300 nm diameter. SPIONs were then precipitated into the shell using the xxx groups to nucleate the particle growth, with the successful incorporation of SPIONs into the shell confirmed by transmission electron microscopy.

Rhodamine B release was then tracked following encapsulation of the core-shell nanoparticles in the *in situ*-gelling zwitterionic hydrogel to immobilize the core-shell nanoparticles at a target therapeutic site. Release was tracked (via fluorescence analysis) into a PBS bath by: (1) heating the nanocomposite hydrogel from 37°C and 45°C; or (2) maintaining the bulk temperature at 37°C but applying alternating magnetic field (AMF) pulses. A 30-minute AMF pulse at 37°C achieved the same release as heating the nanocomposite to 45°C, with the 37°C/No-AMF condition facilitating significantly lower drug release. As such, AMF and temperature can both trigger a significant pulse of drug release from the core-shell nanoparticles.

Conclusion

Core-shell nanoparticles containing a core with a super-physiological glass transition temperature (Tg ~ 45°C), a stabilizing shell, and precipitated superparamagnetic iron nanoparticles in the shell can promote triggered release of drug upon direct heating or an alternating magnetic field stimulus.

Significance

We developed the first polymer-polymer core-shell nanoparticles via aqueous SI-SARA-ATRP and the first polymeric core-shell nanoparticle system used as a drug delivery vehicle, triggerable by either heating or magnetic field activation. We anticipate this system offers potential for facilitating improved pulsatile drug delivery with an improved "off" state release (taking advantage of the very slow drug diffusion through the core particle under glassy conditions), facilitating more specific on-off drug release and thus longer functionality and higher specificity of drug release at a target site.

[1] H. He, et al, "Survey of Clinical Translation of Cancer Nanomedicines-Lessons Learned from Successes and Failures," *Acc. Chem. Res.*, vol. 52, pp. 2445–2461, 2019.

[2] S. S. Said, et al, "Externally Addressable Smart Drug Delivery Vehicles: Current Technologies and Future Directions," *Chem. Mater.*, vol. 31, no. 14, pp. 4971–4989, 2019.

[3] H. Chen, et al, "Endogenous Stimuli-responsive Nanocarriers for Drug Delivery," *Chem. Lett*, vol. 45, pp. 242–249, 2016.

[4] P. J. Rosch, "Electromagnetic Fields: Biological Interactions and Mechanisms, Martin Blank, Editor, American Chemical Society, Washington, D.C., 1995, 550 pgs., $116.95," *Stress Med.*, vol. 13, no. 2, pp. 137–137, 1997.

[5] S. Sabale, et al, "Recent developments in the synthesis, properties, and biomedical applications of core/shell superparamagnetic iron oxide nanoparticles with gold," *Biomater. Sci.*, vol. 5, pp. 2212–2225, 2017.

[6] A. Hervault *et al.*, "Doxorubicin loaded dual pH-and thermo-responsive magnetic nanocarrier for combined magnetic hyperthermia and targeted controlled drug delivery applications," *Nanoscale*, vol. 8, pp. 12152–12161, 2016.

[7] W.-H. Chiang *et al.*, "Superparamagnetic Hollow Hybrid Nanogels as a Potential Guidable Vehicle System of Stimuli-Mediated MR Imaging and Multiple Cancer Therapeutics," *Langmuir*, no. 29, pp. 6434–6443, 2013.

370e Award Submission: Detection and Visualization of HER2 Clustering Using an Aggregation Induced Emission Based Fluorescent Probe Molecule

Yasaman Moradi

The spatio-temporal organization of proteins within the cell membrane can influence basic cell behaviors including signaling and proliferation which play a role in disease initiation and progression. Understanding these control mechanisms can motivate new therapeutic designs. Several methods exist for studying the organization of membrane proteins including electron microscopy, super-resolution optical microscopy methods, and proximity-based assays. However, these methods are either not compatible with living cells or require access to specialized instrumentation. Furthermore, the highest-resolution approaches that can detect the spatial organization of proteins rely on single-molecule imaging methods and cannot be utilized in a high-throughput format.

To overcome many of the limitations of the current methods, we have developed a fluorescent microscopy tool based on a family of fluorophores that exhibit Aggregation Induced Emission (AIE) behavior for studying the spatial organization of membrane-bound proteins. Fluorophores with the AIE property can go through a fluorescence turn-on process when their intramolecular motions are restricted. For example, when AIE molecules form solid aggregates, the emission increases, and this increase in signal can be easily detected using conventional fluorescence microscopy. When proteins cluster or are in close proximity, they behave as a biological mimic for aggregation. Therefore, a tool based on an AIE fluorophore can overcome many limitations of previous methods.

In this work, we demonstrate the development and utilization of an AIE-based tool to study the response of the breast cancer biomarker HER2 to a conventional targeted cancer therapeutic (Trastuzumab). Trastuzumab (Herceptin), an FDA-approved monoclonal antibody therapeutic, is hypothesized to change the biophysical distribution of HER2 in the membrane, disrupting clusters or decreasing densities. The dependence of AIE luminogen's (AIEgen's) fluorescent response on its proximity to other AIEgens can provide insight into this mechanism.

Our AIE-based tool is developed by designing, synthesizing, and characterizing a molecule based on Tetraphenylethylene (TPE) (an AIE fluorophore) and binding it to an antibody specific to the extracellular domain of HER2 through the non-selective NHS Ester-Amine bioconjugation reaction. The target specificity of the AIE-based tool is confirmed through a competing colocalization assay using a HER2+ cell line. Then, the AIE-based tool was utilized for visualizing HER2 clusters in Her2+ cells with the presence of HER2 clusters in comparison to HER2- cells with a more uniform distribution of HER2. Since the molecular motions of the TPE domain of the tool become restricted in the cellular regions where receptors are in proximity of each other, a significantly higher fluorescent signal was observed in the HER2+ cells compared to HER2- cells.

Furthermore, the developed AIE-based tool was utilized in an assay for detecting the effect of Trastuzumab on HER2 clusters. Fluorescein-HER2 antibody, a control that is not sensitive to proximity, and the AIE-based tool are applied to HER2+ cells post-treatment with a range of Trastuzumab concentrations for different time points. The results determine that, unlike the control Fluorescein-HER2 antibody, the AIE-based tool demonstrates a significant reduction in the average fluorescent intensity of the cells treated with higher Trastuzumab concentrations and for longer. This reduction can directly correlate with the disruption of HER2 clusters or the decrease of HER2 densities.

Current results confirm the applicability of the AIE-based tool for visualization of spatial organization of HER2 and set the stage for generalizability of the tool for studying other biological systems with applications in developing improved therapeutics in the future.

370f Award Submission: Covalent Attachment of Horseradish Peroxidase to Single-Walled Carbon Nanotubes for Hydrogen Peroxide Detection

Francis Ledesma

Hydrogen peroxide (H_2O_2), a reactive oxygen species (ROS), is critical to the regulation of mammalian cellular processes such as apoptosis and response to oxidative stress. However, unregulated cellular levels of H_2O_2 leads to erratic DNA mutation and damage, implicating it as a biomarker for diseases such as cancer, Parkinson's, Alzheimer's, and Huntington's. It is thus important to develop sensitive and selective technologies to sense elevated cellular H_2O_2 levels for improved disease diagnostics. Current techniques for H_2O_2 sensing involve chemiluminescent or fluorescent dyes, but these suffer from instability in dynamic environments, photobleaching, and low spatiotemporal resolution.

Here, we developed a nanosensor for detection of H_2O_2. We base our nanosensor on single-walled carbon nanotubes (SWCNTs), owing to their intrinsic near-infrared (nIR) fluorescence, high aspect ratio, and photostability. The selectivity of our nanosensor is conferred by covalently attaching horseradish peroxidase (HRP) to the SWCNT surface, a key enzyme in the oxidative stress response pathway of the horseradish plant. We employed covalent surface modification of SWCNTs with cyanuric chloride, maintaining intrinsic nIR fluorescence[1] while priming the SWCNT surface for further modification with common protein conjugation handles such as carboxylic acid, primary amine, and thiol groups.[2] We functionalized HRP with maleimide groups via NHS-lysine conjugation with a linker named Sulfo-SMCC. We subsequently incubated maleimide-functionalized HRP with thiol-functionalized SWCNTs to covalently attach HRP to SWCNTs without compromising enzymatic activity.

We characterized the nanosensor's physical and optical properties to confirm successful protein conjugation while maintaining intrinsic SWCNT fluorescence and HRP enzymatic activity. Additionally, we assessed the nanosensor's concentration-dependent response to H_2O_2 in solution and demonstrated a limit of detection of 31 uM. Furthermore, we demonstrated the nanosensor's selectivity for H_2O_2 by its lack of response to an analyte panel of related reactive oxygen species and other biologically relevant analytes. Finally, we immobilized the nanosensor on glass to confirm its fluorescent signal is reversible, which we can use to sense elevated peroxide levels in cell lysate from oxidatively stressed MCF-7 cells. Taken together, these results demonstrate covalent attachment of enzymes to SWCNTs with a chemical strategy that preserves both intrinsic SWCNT fluorescence and enzyme function. We anticipate this platform can be adapted to covalently attach other proteins of interest such as other biologically relevant enzymes for sensing and larger protein cargo for cellular delivery including CRISPR-Cas9.

References

1. Setaro, A. *et al.* Preserving π-conjugation in covalently functionalized carbon nanotubes for optoelectronic applications. *Nat. Commun.* **8**, 14281 (2017).
2. Chio, L., Pinals, R. L., Murali, A., Goh, N. S. & Landry, M. P. Covalent Surface Modification Effects on Single-Walled Carbon Nanotubes for Targeted Sensing and Optical Imaging. *Adv. Funct. Mater.* 1910556 (2020) doi:10.1002/adfm.201910556.

370g Award Submission: Desirable Formulations for Inhalable Particulate Immunotherapeutics Using Model Peg-Based Nanoparticle

Emma Sudduth, Catherine Fromen

The pulmonary microenvironment is the first site of infection for numerous pathogens, and as such, is comprised a host of specialized immune cells that are ripe for immunotherapeutic applications. However, determination of design parameters for inhaled nano-immunotherapies to target these cells has yet to be established. As a model platform, poly(ethylene glycol) diacrylate (PEGDA) nanoparticles (NPs) present highly beneficial characteristics for an inhalable formulation as they offer readily tunable physiochemical properties and have shown non-immunogenic effects in the lung. Herein, we characterize the aerosol and lung trafficking properties of model PEGDA NPs for inhalable immunotherapeutic applications through the modification of surface charge and formulation.

Both cationic and anionic formulations of PEGDA NPs were fabricated with similar hydrodynamic diameters around ~225 nm and low polydispersity indexes, through the addition of 10 wt% charge-establishing co-monomers. In liquid form, both formulations demonstrated robust nebulization through a commercial Aeroneb® device via non-significant morphological alterations post-nebulization, ascertained through Cryo-Scanning Electron Microscopy (Cryo-SEM) and ImageJ analysis. Dry-powder formulations of PEGDA NPs were also produced via lyophilization with tunable aerodynamic properties when dispersed through a Plastiape RS01 monodose inhaler. To assess phagocytosis capacity of PEGDA NPs using *in vitro* aerosol delivery, a 3D printed adapter was designed that connected the Aeroneb® device to air-liquid interface (ALI) transwell cultures of innate immune cells. As expected, cationic NPs demonstrated significant uptake at all dosing conditions based on electrostatic charges; however, NP uptake overall by these cells was greatly reduced at ALI, along with different phenotype responses. To corroborate findings from *in vitro* studies, murine studies of lung distribution following orotracheal delivery of cationic and anionic PEGDA NPs were performed using two different age groups. Cellular uptake in immune cell subpopulations, NP trafficking, and the cellular inflammatory profile were compared between the two groups for each of the particle formulations, suggesting distinct design rules for the different age groups. Thus, this work outlines an overall approach to formulate particulate-based inhalable therapeutics that can target specific immune cells in the pulmonary microenvironment for a broad range of inhaled immune engineering applications.

370h Using Nanoparticle-Gated Electrokinetic Membrane Sensors and Janus Diffusivity Sensors for Quantifying Disease-Related Nanovesicles in Biofluids

Sonu Kumar, John Sinclair, Nalin Maniya, Satyajyoti Senapati, Hsueh-Chia Chang

Lipoproteins and Extracellular vesicles (EVs) found in plasma, typically 6-200 nm in size, provide us with information essential for more accurate diagnosis than the current state-of-the-art factors. Diseases such as cancer and cardiovascular diseases, while detrimental, are high preventable, curable, or manageable, provided early screening could be done with > 90% 5-year survival rate in early stage cancer. The current screening process is often invasive, uncomfortable, and potentially harmful and essentially requires whereabouts/location of the tumor. As cancer treatments become more advanced, the need for more robust early screening methods, such as detecting cancer cell biomarkers, becomes apparent. Advancing the field of diagnostic technology has the potential to have profound impact on how we screen for and treat cancer.

Moreover, tumor cells often overexpress specific markers on their surface which aid in tumorigenesis, angiogenesis, and eventually metastasis. Detection of these disease-specific protein biomarkers remains the most promising non-invasive early screening technique. While performing biopsy of such cells lies in the invasive category and requires one to know about the possibility of a tumor or observe some degree of abnormality through commonly used techniques such as mammography and colonoscopy, one can achieve the same using EVs found in plasma that reflect the proteomic signature of their parent cells. A subset of EV know as Exosomes, once considered to be a disposal mechanism for cells, have recently been identified as a strategy to identify and detect the biomarkers of diseased cells. During biogenesis, the exosomes are formed using a portion of the cell's lipid bilayer which is laced with copies of cancer-specific membrane proteins. This is especially important for detecting type of cancer as the exosomes secreted by tumor will contain both the overexpressed and the tissue-specific protein. Additionally, lipoproteins carry surface proteins such as PON1 that provides a significant portion of the atheroprotective properties, and thus prevents cardiovascular disease.

The bottlenecks associated with the measurement of these particles rely on several factors. Firstly, the gold standard for protein measurement, Enzymatic Immunoassays (ELISA), are incompatible with lipids due to redox interference from lipid peroxides on these particles. At the same time, flow cytometry-based methods cannot be extended to such small vesicles. Secondly, ELISA and other commonly used methods lack the sensitivity (>pM) to assess these particles accurately (~fM-pM concentration), requiring more sensitive methods. Thirdly, there can be significant competition from the soluble versions of these proteins or non-targets, which requires us to have a large dynamic range to operate away from sensor saturation. Lastly, the detection method should be fast (<1 hour) and cheap (~$1) so that everyone can afford these tests – perhaps even as an annual or semi-annual test. This would ensure we can detect the disease early, start the treatment early (>85 percent survival for stage I cancer detection) and use these markers to monitor the efficacy of treatment, remission, and recurrence. We have developed Charge-gated Anion Exchange Membrane (AEM)-based Electrokinetic sensors and Janus particle-based diffusometry sensors that are very sensitive (~fM), cheap (~$1), fast (<1 hour), and have a very large dynamic range and is less susceptible to environmental factors.

We have currently validated our strategy of detecting lipid-protein assemblies as well as our technology by assessing cardiovascular risk, detecting glioblastoma and colorectal cancers non-invasively with ongoing studies that include pancreatic, breast and lung cancers. We were able to achieve an impressive AUC ~ 0.99 with PON1-HDL in distinguishing Coronary Artery Disease patients from healthy controls and an AUC > 0.95 for cancer patients (Colorectal and Glioblastoma) from healthy controls. We compared our platform to the state-of-the-art methods used by healthcare providers for the cardiovascular risk assessment, outperforming Cholesterol tests, Apolipoprotein levels, HDL-P levels (AUC~0.65) that accurately predicted CAD in about 60-70% of the patients compared to our platform that predicted >95% of the CAD patients and controls accurately. Identification of cancer and cardiovascular diseases and their associated risks early enough can significantly improve the lives of about 50% of world population that ultimately fall to these diseases. Our platforms can potentially be the annual test for cancer that are highly accurate in diagnosing cancer based on our pilot study of Colorectal Cancer and Glioblastoma.

Publications:

1. Kumar, S.; Maniya, N.; Wang, C.; Senapati, S.; Chang, H.-C., Quantifying PON1 on HDL with nanoparticle-gated electrokinetic membrane sensor for accurate cardiovascular risk assessment. *Nature Communications* **2023,** *14* (1), 557.

220a Investigating Ergodicity-Broken Rotational Dynamics of SWCNT in Hexagonally Packed Colloidal Pores Via Machine Learning

Zhao Tang, Shannon Eichmann, Vida Jamali, Fred MacKintosh, Matteo Pasquali

The diffusion of stiff rods in porous media is not completely understood. Single-molecule imaging of single-walled carbon nanotubes (SWCNTs) in agarose clarified stiff rod diffusion in gels and biological tissues (low volume fraction of flexible obstacles). However, diffusion in porous media with a high volume fraction of hard obstacles—such as packed columns and porous rock—is still incomplete. For example, the impact of the structure of the porous media (pore size, packing type) on stiff rod diffusion is poorly understood. We fabricated porous media from randomly (r-) and hexagonally (h-) packed micron-size colloids to generate sub-micron diameter pores with varying structure and connectivity. We studied the diffusion of SWCNTs (~2 to 10 μm length, ~14 μm persistence length) within the varying pore structures via near-infrared single-molecule microscopy to track SWCNT location, orientation, and curvature. Via automatic image recognition tools, we generated trajectories from SWCNT motion videos, and analyzed the trajectories via machine learning and statistical tools. We find that SWCNT rotation in h-packed colloid pores conforms to a weak ergodicity broken continuous time random walk (CTRW). In h-packed colloidal pores, diffusing SWCNTs face pores with large orientation differences. To enter these pores, SWCNTs must bend significantly, overcoming a large energy barrier, leading to non-convergent averaged waiting times for pore transitions. R-packed colloidal pores have contiguous pores with smaller relative orientational changes; hence, SWCNTs can transition between pores with smaller bending. In these media, we find that slowed SWCNT rotation under highly bent configurations is rare. This shorter, convergent pore transition time leads to diffusion patterns resembling fractional Brownian motion. These findings will help understanding semiflexible sensor dispersion in rock-like porous media.

220b Wetting Transparency of Free-Suspended Single-Layer Graphene on Liquid Substrate

Fan Yang, Annette Thompson, Lei Li

Wetting transparency of single-layer graphene provides immense potential for the development of multi-functional devices in which the wettability of an atomic-thick film can be manipulated by selecting the appropriate supporting substrate. It becomes even more interesting when the substrate is liquid since flowing different liquids beneath the graphene will provide the real-time control of the surface wettability. However, the wetting transparency on liquid substrates has been rarely studied, possibly due to the challenges in performing experiments on liquid substrates. To this end, we have developed a contact angle measurement procedure to determine the surface and interfacial tension of free-suspended single-layer graphene on liquid substrates using Neumann's Triangle Model. By using different testing liquids on top of free-suspended single-layer-graphene or changing liquid substrate underneath single-layer-graphene, our results reveal that single-layer-graphene is nearly complete wetting transparent. This experimental approach can also be applied to other types of 2D materials for similar studies. Our findings provide insight into the underlying mechanism of wetting transparency on single-layer graphene and demonstrate the possibility of controlling the surface wettability of single-layer graphene by engineering the supporting substrate in real time. The results here have important implications in many applications, including water harvesting, microfluidic devices, and liquid-filled single-walled carbon nanotubes.

220c Green Fluorescent Carbon Dots for Sensitive and Selective Detection of Ferrous Ions

Xiangcheng Sun

Iron is critical to living organisms; excess or lack of iron could lead to a series of diseases. In this work, fluorescent carbon dots were developed as an efficient probe for sensitive and selective detection of ferrous ions (Fe^{2+}). We reported on simple and one-pot synthesis of uniform, bright, and green fluorescent carbon dots with aqueous stability in a wide pH range. The formation of green fluorescence is attributed to the doping of nitrogen into carbon materials. The as-synthesized carbon dots demonstrated wavelength-independent emission properties. Fluorescence of carbon dots was quenched efficiently, sensitively and selectively by ferrous ions over a series of other common metal ions, with a detection limit of 16 μM, and a wide detection range of 0.033 -1.044 mM. Efficient quenching of carbon dots could be explained by the efficient binding of ferrous ions on carbon dots surface. Aggregation caused quenching was later proposed and verified as the quenching mechanism of carbon dots toward ferrous ions. The developed carbon dots showed great promise in the development of efficient sensors and adsorbents for ferrous ions.

220d Quantified Carbon Nanotube (n,m) Separation Dependencies in Aqueous Two-Polymer Phase Extraction

Christopher Sims, Jeffrey Fagan

Surfactant controlled aqueous two-polymer phase extraction is a scalable and sensitive method for separating specific (n,m) species of single-walled carbon nanotubes (SWCNTs) to the level of pure left- or right-handed enantiomers. The basis for the separation is progressive isolation of different SWCNT structures into top or bottom self-separating polymer phases until a single structure remains. The selecting mechanism for a top versus bottom phase extraction in a given separation depends sensitively on the type and chemistry of the adsorbed surfactant layer on each (n,m) species, and can be controlled by the choice of competing surfactant(s) and modulator chemicals. In this presentation I will describe the quantified extraction conditions and effects of multiple modulators, such as salts and 3rd component surfactants, to the precise extraction conditions for different (n,m) SWCNTs as determined through high-precision fluorescence measurements. In addition to precise extraction condition data, the effects of surfactant concentration and modulator properties are yielding new mechanistic understanding to the competitive surfactant adsorption at the heart of ATPE separations.

220e Catalyst Aerosol Dynamics in the Floating Catalyst Chemical Vapor Deposition Synthesis of High-Quality Carbon Nanotubes

Arthur Sloan, Muxiao Li, Jui Junnarkar, Glen Irvin, Matteo Pasquali

Engineering materials based on carbon nanotubes (CNTs) have been demonstrated for applications ranging from strong, lightweight structural composites to cables for power and data transmission. Such applications could one day supplant steel, aluminum, and copper in many scenarios. Replacing metals with CNTs could provide a route to reducing the green-house gas emissions and other environmental impacts associated with metal extraction and refining. The success of these applications depends on the supply of long, highly crystalline, CNTs at prices low enough to enable competition with legacy materials. Currently, this is the limiting factor for many of these applications, as suitable CNTs cost more than $2000/kg.

Floating catalyst chemical vapor deposition (FCCVD) is one of the most promising methods for the scalable, continuous synthesis of high-quality CNTs for macroscale applications. FCCVD utilizes an aerosol of catalytic metal particles as active sites for CNT growth. This aerosol may either be preformed from the vaporization and condensation of a metallic source or formed in situ by the decomposition of an organometallic precursor species. Despite the critical importance of the catalyst aerosol to the FCCVD process, this key element remains poorly characterized and poorly understood.

The lack of understanding of this key process has led to an overly simplistic view of the aerosol dynamics within an FCCVD reactor. Critical aspects of the aerosol behavior, such as the potential for the evaporation of the catalyst, have been neglected until very recently. This simplified view has complicated process scale up efforts through traditional geometric similarity methods, and led to an apparent tradeoff between process yield and product quality, contributing to the high prices of high quality CNTs.

The study of the generation and population dynamics of aerosols like the FCCVD catalyst is a well-developed field; however, much of this knowledge has not been applied to the analysis and design of FCCVD processes. In the limited instances where aerosol science was successfully leveraged studies have typically been specific to particular reactors, utilizing computational fluid dynamics (CFD) to understand specific rector configurations and process recipes. The results of these studies were generally not generalizable to the FCCVD process as a whole, limiting their impact and the adoption by the broader community.

In this work we apply proven aerosol modeling techniques to develop a simplified model that captures the key aerosol dynamics present in real reactor. We reduce the reactor to a one-dimensional, plug flow reactor (1D-PFR) and consider the formation of particles by nucleation, the change in their size and concentration by coagulation, condensation, and evaporation, and the loss of catalyst material to the reactor walls. Despite the degree of simplification, the 1D-PFR model agrees well with much more complex, CFD based simulations from the literature while remaining computationally inexpensive and easy to interpret.

We use this model to interrogate the effect of common experimental parameters, such as the reaction temperature, residence time, and injection condition. We consider both the injection of preformed catalyst particles as well as the in situ formation of particles from the decomposition of a precursor species.

The results of these studies reveal that the early phases of particle formation or particle injection are critical to determining the overall dynamics of the process. The simulations show that, in general, in situ particle generation will enable higher particle concentrations due to the diminished role of particle-particle coagulation. Balancing the effects of reaction temperature and residence time is critical for maximizing the particle number concentration when forming the particles in situ. These results agree well with most of the available experimental data and provide more rigorous, aerosol-based explanations for some observations in the FCCVD literature.

220f Macroscopic Weavable Fibers of Carbon Nanotubes with Giant Thermoelectric Power Factor

Natsumi Komatsu, Yota Ichinose, Oliver Dewey, Lauren Taylor, Mitchell Trafford, Yohei Yomogida, Geoff Wehmeyer, Matteo Pasquali, Kazuhiro Yanagi, Junichiro Kono

Thermoelectric materials convert heat into electricity and vice versa, offering great potential for waste heat recovery and solid-state cooling. Carbon nanotubes (CNTs) are promising candidates because of their one-dimensionality along with their unique advantages such as flexibility and light weight. However, preserving the large power factor of individual carbon nanotubes in macroscopic assemblies has been challenging. Here, we studied the thermoelectric properties of aligned CNT fibers with an ultrahigh electrical conductivity ($>10^6$ Sm^{-1}) [1] and observed a giant thermoelectric power factor of 14±5 $mWm^{-1}K^{-2}$ [2]. The observed giant power factor originates from the ultrahigh electrical conductivity achieved through excellent sample morphology, combined with an enhanced Seebeck coefficient through Fermi energy tuning. The unique properties of CNT fibers, weavability, and scalability, combined with the excellent power factor, allowed us to fabricate a textile thermoelectric generator with enough output power to drive a light-emitting diode. Our findings open a route to provide renewable energy production by powerful and weavable thermoelectric generators.

References

[1] L. W. Taylor et al., Carbon, 171, 689 (2021)
[2] N. Komatsu et al., Nat. Commun., 12, 4931 (2021)

220g Thermal Conductivity of Interlayer-Bonded Bilayer Graphene

Afnan Mostafa, Ashwin Ramasubramaniam, Dimitrios Maroudas

Graphene-based metamaterials and derivatives are well-known for their tunable thermomechanical properties through structural modification facilitated by chemical functionalization and defect engineering of graphene. Inserting sp^3-hybridized interlayer C-C bonds in bilayer or few-layer graphene is a type of structural modification that has been pursued successfully for the tunability of thermal and mechanical properties of multi-layered graphene sheets. Examples of such structurally modified few-layer graphene configurations include, but are not limited to, interlayer-bonded twisted bilayer graphene (IB-TBG), interlayer-bonded graphene bilayers with randomly distributed individual interlayer bonds (RD-IBGs), and two-dimensional (2D) diamond or diamane consisting of fully interlayer-bonded graphene bilayers. 2D diamond superstructures in IB-TBG with the periodicity of the Moiré superlattice pattern of twisted bilayer graphene (TBG) are generated from TBG through insertion of interlayer C-C bonds as a result of patterned hydrogenation of each graphene layer [1,2]. If the chemical functionalization of the graphene layers is not patterned, interlayer C–C bonding does not occur in a periodic arrangement and, instead of 2D diamond superstructures, RD-IBGs are formed. In general, such interlayer-bonded bilayer graphene sheets are expected to be uniquely promising materials, which combine the exceptional properties of graphene and diamond.

To this end, here, we report results from a systematic analysis of thermal transport in interlayer-bonded bilayer graphene sheets based on molecular-dynamics (MD) simulations. We find that the introduction of interlayer C-C bonds in IB-TBG 2D diamond superstructures causes an abrupt drop in the thermal conductivity of pristine, non-interlayer-bonded bilayer graphene, while further increase in the interlayer C-C bond density (2D diamond fraction) leads to a monotonic increase in the thermal conductivity of the resulting superstructures with increasing 2D diamond fraction toward the high thermal conductivity of 2D diamond (diamane) [3]. We also find that a similar trend is exhibited in the thermal conductivity of RD-IBGs as a function of interlayer C-C bond density, but with the thermal conductivity of the IB-TBG 2D diamond superstructures consistently exceeding that of RD-IBGs at a given interlayer bond density [3]. We analyze the MD simulation results employing effective medium and percolation theories and explain the predicted thermal conductivity dependence on interlayer bond density on the basis of lattice distortions induced in the bilayer structures as a result of interlayer bonding [3]. Our findings demonstrate that the thermal conductivity of IB-TBG 2D diamond superstructures and RD-IBGs can be precisely tuned by controlling interlayer C-C bond density and have important implications for the thermal management applications of interlayer-bonded few-layer graphene derivatives, including their use as heat sinks in nanoelectronic devices.

[1] A. S. Machado, D. Maroudas, and A. R. Muniz, Appl. Phys. Lett. **103**, 013113 (2013).

[2] M. Chen, A. R. Muniz, and D. Maroudas, ACS Appl. Mater. Interfaces **10**, 28898–28908 (2018).

[3] A. Mostafa, A. Ramasubramaniam, and D. Maroudas, Appl. Phys. Lett. **122**, 133101 (2023).

281a Controlled Synthesis of Carbon Nanoparticles from Lignin Using Single-Step Aerosol Methods for Application in Energy Storage

Sujit Modi, Onochie Okonkwo, Sulay Saha, Hao Zhou, Shalinee Kavadiya, Marcus Foston, Pratim Biswas

With the rapidly growing demand for renewable energy and portable devices, developing high-performance energy storage materials and meeting sustainability requirements at the same time is of great significance. Lignin, a biomass constituent and byproduct waste of biorefineries, is one of the few renewable sources of carbon that have the abundance and desirable properties to displace or replace fossil-derived carbon sources. A renewable resource, low cost, and high aromatic content make lignin an ideal precursor to obtaining carbon materials for energy storage. However, present efforts on the large-scale synthesis of carbon materials from lignin are limited either by multistep and batch processes or the use of activating agents/templates [1].

The present work aims to engineer a process that valorizes waste and by-product lignin. In particular, we develop an original and innovative method, based on a one-step continuous aerosol technique, to manufacture high-performance carbon nanoparticles for energy storage without the use of activating agent/template. This one-step approach requires a significantly lower time for synthesis: an order of seconds in comparison to hours for existing methods. Temperature and residence time are expected to have a key impact on the properties of carbon materials synthesized from lignin, hence, an aerosol technique utilizing a furnace aerosol reactor is used; furnace aerosol reactor is known for enabling precise control over a wide range of temperatures and residence times [2]. Subsequently, the effect of temperature and residence time inside the reactor on carbon nanoparticle size, functional groups, surface area, and morphology is systematically investigated. Furthermore, the as-obtained carbon nanoparticles are tested for specific capacitance and the best-performing material (surface area 925 m^2/g) exhibited a specific capacitance of 247 F/g at 0.5 A/g with excellent capacity retainment of over 98 % after 10,000 cycles. This is a clear demonstration of their superior performance compared to supercapacitors synthesized earlier from lignin.

Moreover, the size and morphology of carbon particles, which are impacted by interparticle collision and sintering, are known to play a crucial role in their stability as well as energy storage performance. However, experimental as well as modeling efforts on size and morphology-controlled synthesis of carbon nanoparticles from lignin are limited, partly due to the complexity of studying collision and sintering under simultaneous lignin reactions (pyrolysis). We systematically investigate the kinetics of reaction, sintering, and collisions using a novel and generalizable numerical model based on a geometric modeling approach [3]. Knowledge of kinetics will provide better control over the size and morphology for efficient utilization of carbon nanoparticles in energy storage.

Overall, the simple (one-step, continuous, and rapid) operation, avoiding the use of activating/templating chemicals, and precise control over size and morphology make the aerosol technique a promising candidate for the scalable and sustainable synthesis of energy storage materials.

References:

[1] J.W. Jeon, L. Zhang, J.L. Lutkenhaus, D.D. Laskar, J.P. Lemmon, D. Choi, M.I. Nandasiri, A. Hashmi, J. Xu, R.K. Motkuri, Controlling porosity in lignin-derived nanoporous carbon for supercapacitor applications, ChemSusChem 8(3) (2015) 428-432.

[2] S. Modi, M.B. Foston, P. Biswas, Controlled Synthesis of Smaller than 100 nm Lignin Nanoparticles in a Furnace Aerosol Reactor, ACS ES&T Engineering (2023).

[3] S. Modi, O. Okonkwo, H. Zhou, S. Kavadiya, M. Foston, P. Biswas, Geometric Model for Predicting the Size and Morphology Evolution of Multiparticle Aggregates during Simultaneous Reaction and Sintering, Chem. Eng. J. (2023) 141423.

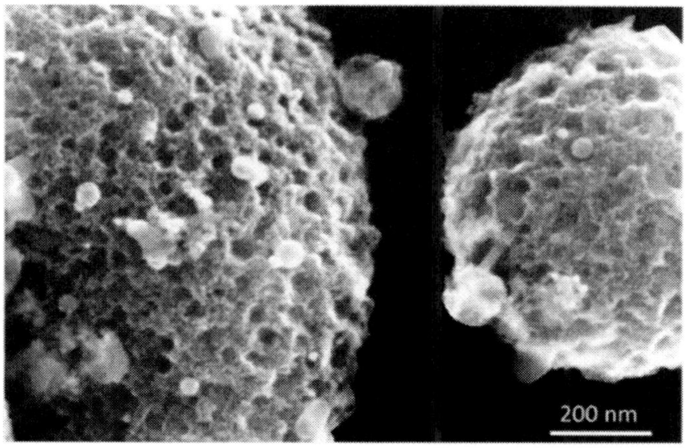

Lignin derived high surface area
carbon nanoparticles (~ 1000 m^2/g)

281b Obtaining Structural Information of Carbon Black in Complex Mediums for Energy Storage Systems with Rheo-Electric Measurements

Qingsong Liu, Yoshita Gupta, Jeffrey Richards

Carbon black (CB) serves as an important carbon nanomaterial given its industrial applications such as fuel cell catalyst inks, flow battery conductive additives, and lithium-ion battery slurries. It is known that when formulated into suspensions, CB shows complicated rheological and electrical behaviors originating from evolving microstructure of CB particles and both the rheological and electric properties are important to engineer and optimize the performance of energy storage systems. While much is known about CB both at rest and under flow in simple mediums, the rheology and the electric properties of CB in complex mediums need to be investigated as CB suspensions are usually added with other components when formulated into industrial products.

Specifically, slurries of carbon black used in lithium-ion battery manufacturing incorporate dissolved polymers and active materials. We used rheo-electric measurements first to show that the addition of polymers modifies the structure of CB agglomerates through the indirect effect on the solvent viscosity. Simultaneous electric measurements were conducted on these CB/polymer suspensions in shear flow as a function of carbon and polymer composition. We found that the CB agglomerate size information can be derived from normalized suspension viscosity through rheological measurements. Additionally, the further validation of such size information is evident in the dielectric strength of the relaxation process obtained from the electric measurements. We further conducted rheo-electric measurements on lithium-ion battery slurries which contain CB, polymers, and active materials. From the dielectric spectrums, we obtained the structure and the dispersion of CB within the battery slurries, which were shown to be strongly correlated with the measured rheology.

These findings are valuable to energy storage systems containing CB as the dispersion of CB can be predicted and controlled with tunable processing conditions to optimize the performance. In addition, the unique insight gained from the rheo-electric measurements can be further extended to engineer the next-generation of CB nanomaterials for energy storage systems.

281c Investigation of the Electrical Properties of Carbon Black Suspensions Undergoing Shear-Induced Microstructural Rearrangement

Paolo Ramos, Julie Hipp, Qingsong Liu, Jeffrey Richards

Carbon black is ubiquitous as a conductive additive in electrochemically active slurries for emergent energy storage and capacitive desalination systems. When suspended, these high-structured, electrically conductive particles agglomerate, creating a charge-carrying microstructure that facilitates electron transport. Furthermore, this microstructure evolves in response to flow, leading to a varied electrical response depending on the suspending medium and the flow conditions. To rationalize this varied response, we dispersed Vulcan XC72R carbon black particles in Newtonian fluids with varying dielectric constants and viscosities. The formulated carbon black suspensions also spanned a range of volume fractions above the electrical and mechanical percolation thresholds for each solvent. We sheared the suspensions from 2500 to 10 s^{-1} and characterized the microstructural rearrangement of the agglomerates by calculating the Mason number. At each flow regime, we performed in situ impedance frequency sweeps on the suspensions and obtained their conductivities, relative permittivity, and dielectric strengths by fitting the dielectric spectra to several models. By parametrically pairing the measured mechanical and dielectric properties, we reveal the systematic dependence of the mechanism of charge transport on not only the microstructure of the carbon black agglomerates but also the solvent properties. We believe that this qualitative insight will provide crucial design rules towards the development of slurries with optimal performance.

281d Polarization Resolved Fluorescence Detection of Carbon Nanotube Corona Phases

Xiaojia Jin, Matthias Kuehne, Xun Gong, Amandine Ucko, Volodymyr Koman, Michael Strano

Circular dichroism measures how chiral particles absorb right- and left- circularly polarized light differently to distinguish pairs of enantiomers. It can also elucidate differential photoabsorption of chiral fluorescent species via their emissions. However, this method has generally been restricted to bulk samples. Hence, these methods conceal information on the specific chiroptical response of a single particle. Furthermore, circular dichroism has been restricted in time- resolution as well as spectral range. In this work, we observed fluorescence detected circular dichroism (FDCD) of individually deposited single walled carbon nanotube (SWNT) corona. A custom-built setup compromising a linear polarizer (LP) and a quarter waveplate (QWP) was used to induce circularly polarized light (CPL). The light polarization state at the sample plane was analyzed using Stoke's parameter. Experimental results showed that linear dichroism of the dichroic filter inside the microscope changes the polarization state of the incoming beam drastically, and thus inducing elliptically polarized light at the sample plane. A liquid crystal variable retarded (LCVR) was used to correct the polarization state of light and yield CPL at the sample plane. Singly-dispersed SWNT with different wrapping polymers were deposited onto a glass substrate. Alternating cycles of right- and left-handed circularly polarized light was delivered to excite the sample. The fluorescence intensities of each individual SWNT were tracked. The dissymmetry (g) parameters $g = 2*(I_L-I_R)/(I_L+I_R)$ (where I_L and I_R are the measured fluorescence intensities with left and right circularly polarized light, respectively) were analyzed from the fluorescent signals to understand photoabsorption differences. A non-zero value of dissymmetry parameter will provide information on how the corona modulates SWNT photoluminescence, an outcome not observable by bulk characterization methods. This observation can potentially enable the measurement of chiral species local interactions without a separations process.

244a Investigation of Advanced Catalysts and Metal Oxide Supports Having Nanoporous Structures Towards Oxygen Evolution Reaction in Proton Exchange Membrane Water Electrolyzer

Chaekyung Baik, Je Yeon Choi, Hyung Joo Lee, Chanho Pak

Recently, the proton exchange membrane water electrolysis (PEMWE) which can store excess electricity as a form of transportable hydrogen has been attracting great attention as a power-to-gas technology for sector coupling of renewable power and industrial, transportation, and heating [1]. Owing to the sluggish oxygen evolution reaction (OER) that can generate the oxygen and proton from the water and harsh reaction conditions like low pH and high voltage in the anode of the PEMWE, the precious metal oxide catalysts such as IrO_2 and RuO_2 should be applied in the membrane electrode assembly (MEA). However, there is a challenge to decrease the amount of the noble metal to the level of 0.01 g_{Ir}/kW [2] in the MEA to commercialize PEMWE at a GW scale. Thus, the catalysts for the OER, which are mainly iridium(Ir)-based composition, have been developed for improving the activity and stability through several approaches such as control of oxidation states [3], supported catalysts [4], and mixed metal oxides [5].

In this presentation, the effect of the porous structure generated inside the OER catalyst and the metal oxide support on the activity toward OER will be discussed. To produce a nanoporous structure, hard templates such as an ordered mesoporous silica (OMS) [5], a silica nanosphere [6], and a polystyrene nanosphere are adapted. For example, a three-dimensional mesoporous IrO_2/RuO_2 catalyst was prepared by sequential nano-replication using OMS and the Adams method [5]. The effects of the structural changes of the binary oxide catalyst on the OER activity were analyzed. As the second example, to generate the concave pore structure inside the OER catalyst, silica nanospheres of different sizes were used to induce the formation of the larger mesopores over 10 nm and macropores. The macarpore-dominat OER catalyst showed improved activity and stability. In addition, the generation of the porous structure inside the metal oxide support will be discussed to emphasize the benefit of this approach.

[1] IRENA "Innovation Landscape Brief: Renewable Power-to-Hydrogen", International Renewable Energy Agency, Abu Dhabi, 2019 at www.irena.org/publications.

[2] M. Bernt, A. Siebel, H. A. Gasteiger, *J. Electrochem. Soc.,* **165** (2018) F305.

[3] S. W. Lee, C. Baik, D.-H. Kim, C. Pak, *J. Power Sources*, **493** (2021) 229689.

[4] S. Zhao, A. Stocks, B. Rasimick, K. More, H. Xu, *J. Electrochem. Soc.*, **165** (2018) F82.

[5] S. W. Lee, C. Baik, T.-Y. Kim, C. Pak, *Catal. Today*, **352** (2020) 39.

[6] C. Baik, S. W. Lee, C. Pak, *Micropo. Mesopo. Mater.*, **309** (2020) 110567.

244b Mechanistic Role of Cations in the Electric Double Layer during Electrocatalysis

Hyungjun Kim

Electrocatalysis, whose reaction venue locates at the catalyst–electrolyte interface, is controlled by the electron transfer across the electric double layer, envisaging a mechanistic link between the electron transfer (ET) rate and the electric double layer structure. A fine example is so-called a "*cation effect*", which stands for a strong dependence of electrocatalytic activity on the alkali metal cation (M^+) identity. Although the cation effect has been observed in a variety of electrochemical reactions, there is yet to be a unified molecular picture for that. To address this issue, our group has been developing a quantum-mechanics-based multiscale simulation method, which can elucidate the M^+-coupling capability to possible intermediates. In this talk, I will discuss our atomic and electronic level understandings on the cation effect based on our new computational capability. Particularly, we theoretically establish H^+- and M^+-associated ET mechanisms; the former is often termed as a proton-coupled ET (PCET), and the latter can be analogically termed to be a cation-coupled ET (CCET). I further suggest an experimental route to prove the CCET-based mechanisms, and how to utilize the CCET pathways to tailor the electrocatalytic activity.

244c Facile Defect Engineering to Tailor Porous Coordination Polymers for Energy-Efficient Gas Separation

Jong Suk Lee

Membrane-based separation technology, a pressure-driven process, is more energy efficient than thermally-driven separation processes because it does not involve a phase change during separation. Zeolitic imidazolate frameworks (ZIFs) are attractive porous coordination polymers for gas separation composed of tetrahedrally-coordinated transition metal ions and imidazole organic ligands. They can be incorporated into a polymer matrix in the form of dispersed nanofillers, called mixed matrix membranes (MMMs), to provide potential scalability and superior molecular size-sieving ability. Recently, our group reported a new facile method of fabricating highly processable ZIF nanofillers, a so-called defect engineering approach, by using monodentate alkylamine modulators. It effectively suppresses the formation of non-selective interfacial voids around nanofillers in MMM and enhances the molecular sieving ability of ZIF-8 for C_3H_6/C_3H_8 separation. In addition, it can be extended to different gas separations, including CO_2/N_2 (or CO_2/CH_4) and C_2H_4/C_2H_6 separations, via either post-synthetic modification (PSM) or in-situ solution-based synthesis. In this talk, I will discuss engineering the desired molecular structure of ZIF nanofillers via defect engineering for gas separation as well as transforming a flat sheet MMM into a hollow fiber configuration.

244d Nanophase-Segregated Structures and Transport in Aquivion Membrane for Polymer Electrolyte Membrane Fuel Cell: Molecular Modeling Approach

Mohammed Bazaid, Seung Soon Jang

In this study, molecular dynamic simulation method is employed to investigate the structure and transport properties of polymer membrane (Aquivion) consisting of perfluorinated polymers with short side chain in comparison to Nafion. From experimental and computational work, there is a general consensus that the favored properties of polymer electrolyte membrane (PEM) system are a result of nano-phase segregation of hydrophilic clusters in a hydrophobic matrix. Aquivion has a similar chemical structure to Nafion with hydrophobic backbone and hydrophilic sulfonate group. By performing MD simulations of Aquivion and Nafion under the same conditions such as temperature and water content, we compared the nanophase-segregation, transport of water and proton, local structures in the membranes as a function of temperature and water content.

244e First-Principles Investigation of Band-Gap Tuning in 2D Hybrid Organic-Inorganic Perovskites

Omar Allam, Yoonseo Nah, Ilgeum Lee, Dong Ha Kim, Seung Soon Jang

Hybrid organic-inorganic perovskites (HOIPs) have garnered considerable interest for various applications, owing to their exceptional electrical and optical properties such as high photoluminescence quantum efficiency and bright emission. This study utilizes density-functional theory (DFT) methods to examine structure-bandgap relationships in 2D HOIPs with the goal of fine-tuning their properties for diverse optoelectronic applications. In particular, the impact of spacer size in the interlayer region of 2D HOIPs remains unclear. Our DFT analysis reveals a crucial correlation between the steric size of organic spacer moieties and the perovskite bandgap. Although perovskite bandgaps can be adjusted by mixing halides, providing precise tunability of optical properties through controlled halide composition, mixed halide perovskites exhibit detrimental spectral instability, hindering their commercial use in LED applications. Using the nudged elastic band (NEB) method, our DFT calculations uncover the root cause of this spectral instability: the redistribution of the mixed halides. Our mechanistic investigation with DFT offers critical insights into achieving bandgap tuning in HOIPs and sheds light on the fundamental factors behind the observed spectral instability in devices containing mixed halide compositions.

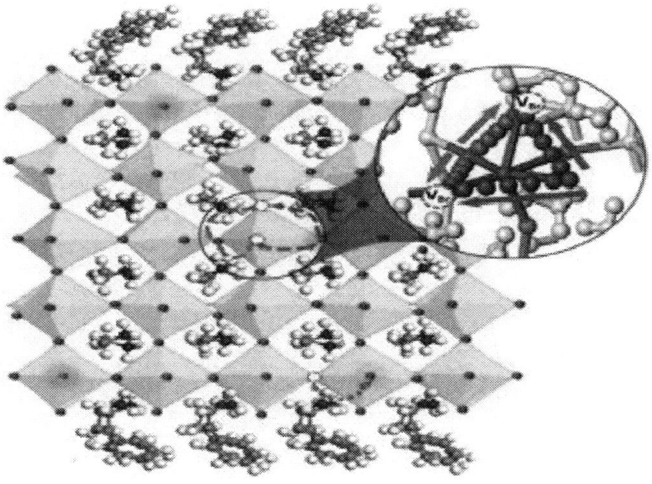

244g Photothermal Effects of Plasmonic Nanoparticles to Enhance Diels–Alder (DA) Reactions in Reversible Thermosets

Arnob Dipta Saha, Brandon McReynolds, Madeline Finale, Youngmin Lee, John McCoy, Sanchari Chowdhury

Thermosets' excellent mechanical properties, chemical resistance, and thermal stability make them an appealing material for additive manufacturing. Irreversible crosslinking and slow curing rate of thermosets like epoxy, however, makes layer by layer printing difficult. We have developed a reversible
thermosets polymer incorporating thermo-reversible covalent adaptable network (Diels–Alder (DA) reactions) into it and mixing that with refractory plasmonic titanium nitride (TiN) nanomaterials. TiN nanoparticles can efficiently, rapidly convert broad spectrum visible light into localized heating to trigger
rapid polymerization and depolymerization reactions making them suitable for reprocessing as well as additive manufacturing. Additionally, it provides opportunity for targeted repair of defect on a 3-D printed epoxy structure and smoothing the rough surface of the finished products. In this study, a comparative investigation was conducted between heat and light stimulus to examine their effect on Diels-Alder reactions. Our in-situ FTIR study of reversible epoxy/TiN composites under light suggested
that broad spectrum white light can significantly enhance the reversible reactions. Interestingly, the reaction rate with light is significantly faster than the reaction driven by heat even though the average bulk temperature of the samples is similar. Plasmonic TiN nanostructures concentrate incident light in nanometer-sized volumes to generate nanoscale heating which cause the local temperature around nanoparticles hotter than the bulk temperature measured. Moreover, optical microscopy and atomic force microscopy images demonstrate that we can use light to precisely depolymerize a selected surface
region of manufactured items to smooth out their rough surfaces or repair any cracks.

244h Thermoelectric Transport Coefficients for Weyl Semimetals in the Family of Transition Metal Monopnictides with Dislocation Deffects

Enrique Munoz

Weyl semimetals (WSMs) constitute a remarkable example of three-dimensional, gapless materials with nontrivial topological properties. First proposed theoretically and, more recently, discovered experimentally on TaAs crystals [1] and other transition metal monopnictides. In a WSM, the band structure possesses an even number of Weyl nodes with linear dispersion, where the conduction and valence bands touch. These nodes are monopolar sources of Berry curvature, and hence are protected from being gapped since their charge (chirality) is a topological invariant [2]. In the vicinity of these nodes, conducting states behave as Weyl fermions, i.e., massless quasi-particles with pseudo-relativistic Dirac linear dispersion [2]. In Weyl fermions, conserved chirality determines the projection of spin over their momentum direction, a condition referred to as "spin-momentum locking".

The presence of Weyl nodes in the bulk electronic spectrum determines the emergence of Fermi arcs [1,2], the chiral anomaly, and the chiral magnetic effect, among other remarkable properties [2]. Therefore, considerable attention has been paid to understand the electronic transport properties of WSMs [2]. Much less explored are the consequences of mechanical strain and defects in WSMs. From a theoretical perspective, it has been proposed that different types of elastic strains can be modeled as gauge fields in WSMs [3-6]. In previous works, we have studied the combined effects of a single torsional dislocation and an external magnetic field on the electronic [4,6] and thermoelectric [3,5] transport properties of WSMs, using the Landauer ballistic formalism in combination with a mathematical analysis for the quantum mechanical scattering cross-sections [3-5].

In this work, we extend our previous analysis to study the case of a diluted, uniform concentration of torsional dislocations and its effects on the electrical conductivity of type I WSMs. In contrast to our former studies [3-5], here we employ the Kubo linear-response formalism at finite temperatures. This requires explicitly calculating the Green's functions for the system, including the multiple scattering events due to the random distribution of dislocation defects. For this purpose, we first analyze the scattering phase shift arising from a single torsional dislocation, and then we obtain the corresponding (retarded and advanced) Green's function in terms of the T-matrix elements by solving analytically the Lippmann–Schwinger equation. We further extend this analysis, by incorporating the effect of a random distribution of such dislocations, with a fixed concentration n_d. Finally, we analyze the correction due to the scattering vertex, and by including this additional contribution, we calculate the electrical conductivity, thermal conductivity and Seebeck coefficient from the Kubo formalism, as a function of temperature and concentration of dislocations. We present explicit evaluations of our analytical expressions for the thermoelectric transport coefficients as a function of temperature and concentration of dislocations n_d, for several materials in the family of transition metals monopnictides, i.e., TaAs, TaP, NbAs and NbP, where the corresponding microscopic parameters, estimated by ab initio methods.

References

[1] Xu, S.Y.; et al., Science (2015), 349, 613–617.

[2] Burkov, A., Annu. Rev. Condens. Matter Phys. (2018), 9, 359–378.

[3] Muñoz, E.; Soto-Garrido, R., J. Appl. Phys. (2019), 125, 082507.

[4] Soto-Garrido, R.; Muñoz, E.; Juricic, V., Phys. Rev. Res. (2020), 2, 012043.

[5] Bonilla, D.; Muñoz, E.; Soto-Garrido, R., Nanomaterials (2021), 11, 2972.

[6] Bonilla, D; Muñoz, E., Nanomaterials (2022) 12, 3711.

The family of transition metal monopnictides

$a = 15\,nm$

D. A. Bonilla and E. Muñoz, Nanomaterials 12 (20), 3727 (2022)

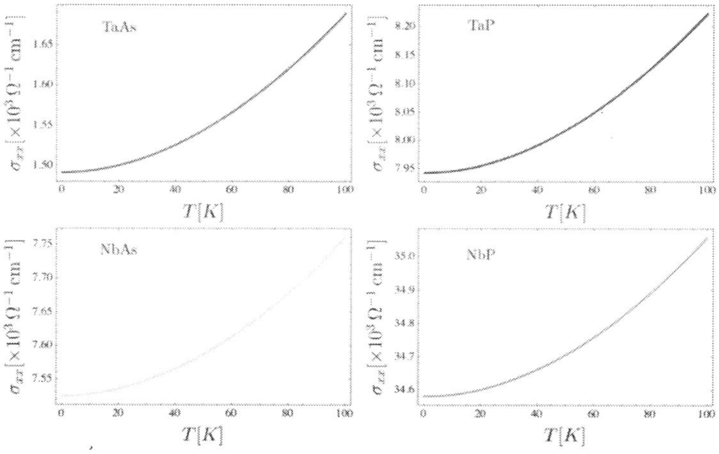

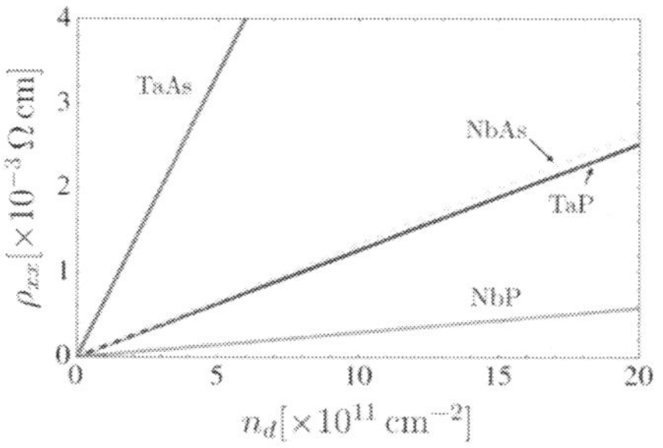

Material	$v_{F,x}^{(+,+)}$	$v_{F,x}^{(-,+)}$	$v_{F,x}^{(+,-)}$	$v_{F,x}^{(-,-)}$
TaAs	3.2	-5.3	2.6	-4.3
TaP	3.7	-5.4	2.0	-3.9
NbAs	3.0	-4.8	2.5	-3.2
NbP	3.0	-5.1	1.7	-2.4

$n_d = 10^{11}\,cm^{-2}$

$\theta = 15^o$

$\alpha = \frac{3\pi}{4}$

Material	τ [10^{-13} s]	τ_{tr} [10^{-13} s]
TaAs	2.2	2.6
TaP	2.4	3.2
NbAs	2.2	3.1
NbP	2.4	4.2

Material	σ_0 [$10^3\ \Omega^{-1}$ cm^{-1}]
TaAs	1.5
TaP	7.9
NbAs	7.5
NbP	34.6

Fermi velocities from ab-initio band structure calculations

D. Grassano et al., Scientific Reports 8, 3534 (2018)

300a Molecular Design of Polymer for Solid State Electrolyte Application in Li-Ion Battery: Molecular Modeling Approach

Omar Allam, Seung Soon Jang

Solid polymer electrolytes (SPEs) have gained considerable attention recently, offering key advantages over liquid electrolytes, such as nonflammability, improved mechanical stability, and minimized electrode material dissolution. In this study, we employ a molecular modeling approach consisting of density functional theory (DFT) and molecular dynamics (MD) simulations to achieve a fundamental understanding of the relationship between the molecular design of polymer and the properties/performances of Li-ion battery application. For this purpose, we investigate the structural and transport properties of carbonate-based SPEs compared to widely used liquid carbonate electrolytes. Specifically, we examine the influence of carbonate pendant group composition on the nanophase morphology of amorphous polycarbonates and Li-ion solvation transport, contrasting them with liquid electrolyte systems like dimethyl carbonate and ethylene carbonate. Furthermore, we investigate how linear versus cyclic carbonate sidechains affect the polymer chain relaxation and the combination of different carbonate pendant group types. This research offers valuable insights for determining the optimal design of polymers to achieve a favorable nanophase morphology and dynamics, enhancing ion transport and mitigating the dissolution of organic cathode materials in the electrolyte.

300b Unravelling Ultrafast Li-Ion Transport in Functionalized Metal-Organic-Framework Based Battery Electrolytes

Zheng Chen

Non-aqueous fluidic transport and ion solvation properties under nanoscale confinement are poorly understood, especially in ion conduction for energy storage and conversion systems. Herein, metal-organic frameworks (MOFs) and non-protic electrolytes are studied as a robust platform for developing a molecular-level understanding of electrolyte behaviors in confined spaces. By employing computer simulations, along with spectroscopic and electrochemical measurements, we demonstrate several phenomena that deviate from the bulk, including modulated solvent molecular configurations, aggregated solvation structures, and tunable transport mechanisms from quasi-solid to quasi-liquid in functionalized MOF structures. Technologically, taking advantage of these confinement effects may prove useful for addressing stability concerns associated with volatile organic electrolytes while simultaneously endowing ultra-fast transport of solvates, resulting in improved battery performance, even at extreme temperatures. The molecular-level insights presented here further our understanding of structure-property relationships of complex fluids at the nanoscale, information that can be exploited for the predictive design of more efficient electrochemical systems.

300c *Computational Design of High Performance Liquid Organic Hydrogen Carriers and Their Corresponding Dehydrogenation Catalysts*

Jeong Woo Han

The use of liquid organic hydrogen carrier (LOHC) has become an attractive method for hydrogen storage due to its ability to store and release hydrogen using reversible hydrogenation and dehydrogenation. For a substance to be an effective LOHC material, it needs to meet certain criteria such as having both a sufficient dehydrogenation enthalpy and hydrogen storage capacity. In this presentation, we will discuss our efforts to identify potential LOHC candidates through a combination of high-throughput computational screening and machine learning. Additionally, we will explore the mechanism behind the catalytic dehydrogenation of dicyclohexylmethane derivatives on Pd and Pt catalysts, which are promising candidates for LOHC. Our findings will provide valuable insights for designing high-performance LOHC materials and their dehydrogenation catalysts.

300d Operando Study Electronic and Magnetic Transitions in Lithium and Sodium Batteries Using X-Ray Emission/Absorption Spectroscopy

Faisal M. Alamgir

Tracking the change in electronic and local structure is crucial to investigate and improve the stability and performance of rechargeable batteries. Doing so under *operando* conditions is challenging but, if done correctly and complementarily with *ex-situ* studies, can reveal the intimate connections between local point-group structure/symmetry, the oxidation states of the constituent elements and their spin states. We will discuss the advances in both lab-based and synchrotron-based x-ray absorption and high-energy-resolution x-ray emission spectroscopy that we used to study not just the internal reactions within batteries during charge and discharge, but also the intrinsic property changes (e.g. magnetic) induced in the active materials by these reactions. The cases that will be examined are lithium and sodium iron phosphates (LFP/NFP), lithium nickel manganese cobalt oxide of various stoichiometries (NMC 622 and NMC 811), iron oxide (Fe_2O_3), and lithium cobalt oxide (LCO). As NMC and LCO are structural siblings, we will specifically look at the connections and subtle differences between them as they undergo electrochemically induced chemical and physical changes.

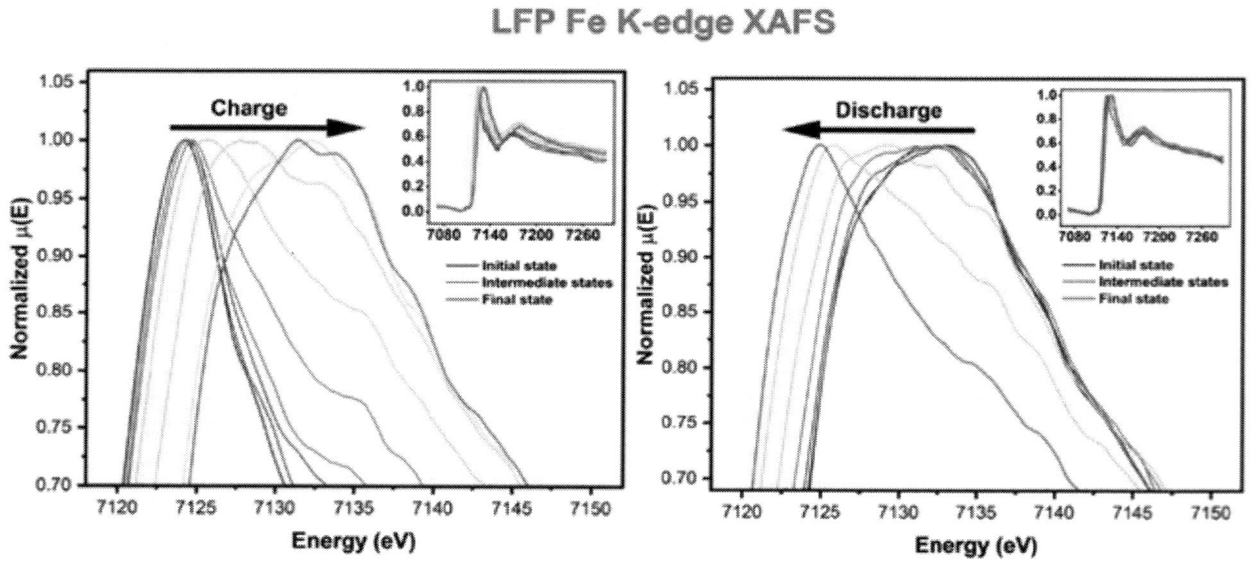

300e Developing Non-Aqueous and Low-Temperature Iron Batteries with V_2O_5 Nanowire Cathodes

Raju Vadthya, Hao Nguyen, Matt Powell, Shuya Wei

Batteries remain as one of the promising technologies to store and deploy energy upon demand. The most successful Li-ion technology is limited for large-scale application due to the safety challenges probed using the organic solvents. However, multivalent metal-based batteries such as Mg, Al, Zn, Ca *etc.* remain as promising and potential alternatives to the Li-ion technology due to their huge abundance in the earth's crust, multiple electrons transfer upon redox reactions, ambient in the environment, non-toxic and remarkably high theoretical capacity. Another type of multivalent metal, Iron, is often overlooked as an electrode for energy storage applications due to the positive standard redox potential value (-0.44V). However, its huge presence in the earth's crust (5.63 wt% vs. 0.002wt% for Li), ability to exist in eight oxidations states (i.e. -2 to +6), and significantly lower cost (0.06 $USD.kg^{-1}$ vs. 19.2 $USD.kg^{-1}$ for Li) remain significant for the possibility to explore Iron as an potential electrode.

Deep eutectic electrolytes (DEEs) are analogues of ionic-liquids and offers appreciably high ionic conductivity, wide thermal and potential window, low volatility, facile preparation, low-cost, non-toxicity *etc.* Here, we report a non-aqueous and DEE based on triethylamine hydrochloride (TEAHCl) with Iron(III) chloride ($FeCl_3$) for the application of rechargeable Iron batteries. The depression of freezing point for the 1.7:1 molar ratio containing TEAHCl:$FeCl_3$ is identified as <0 °C through the differential scanning calorimetry. The structural property correlation to the ionic complexations formed in DEEs were thoroughly investigated using the Fourier Transform Infrared and Raman spectroscopies which concludes the presence of $FeCl_4^-$ and $Fe_2Cl_7^-$. The DEEs show appreciably high ionic conductivity of ~1.2 $mS.cm^{-1}$ and ~100% efficiency of plating and stripping in a symmetric Fe||Fe cell, sustaining for >400 hours. V_2O_5 nanowires, synthesized via hydrothermal routes, used as cathode against Fe anode. Fe||1.7:1 Et_3NHCl:$FeCl_3$||V_2O_5 cell showed open circuit potential of ~1.0V. The galvanostatic cycling in the potential window of 0.0-2.5V *vs.* Fe/Fe^{+3} retrieves a high capacity of ~80 mAh/g. The charge storage mechanism of the cell is thoroughly delineated by imparting XRD and XPS analysis on the cycled electrodes. Our research effort to design a non-aqueous electrolyte for Iron batteries certainly assist to gain the characteristic properties of Iron as an electrode and helps to formulate a strong working hypothesis in the future.

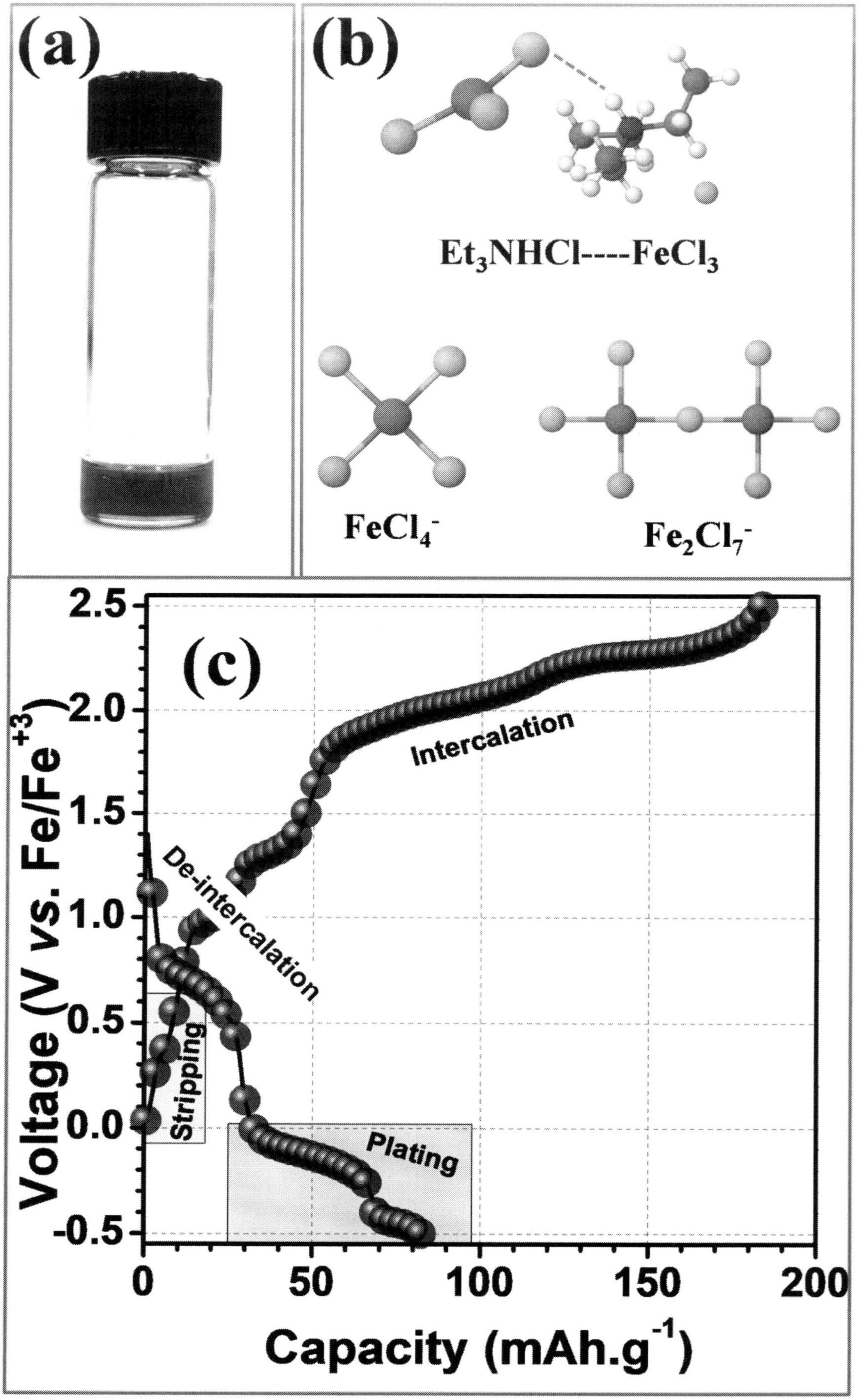
(a)
(b)
Et$_3$NHCl----FeCl$_3$
FeCl$_4^-$
Fe$_2$Cl$_7^-$
(c)
Voltage (V vs. Fe/Fe^{+3})
Capacity (mAh.g^{-1})
Intercalation
De-intercalation
Stripping
Plating
2.5
2.0
1.5
1.0
0.5
0.0
-0.5
0
50
100
150
200

300f Conductive Membrane Coating for High Performance Vanadium Redox Flow Battery Achieved By Air-Controlled Electrospray

Yiqi Shao, Xiaosi Gao, Shuo Jin, Yong Lak Joo

With a growing need for the development of safe, large-scale energy storage technologies, redox flow batteries known for their high capacity, energy efficiency and low cost have attracted huge attention. Among all kinds of different flow batteries, Vanadium flow redox batteries (VFRB) are the most promising system. Yet, still, VFRB possess many issues to be solved. One of the most critical issues is the crossover of ions through expensive Nafion membrane. During the charge and discharge process, Vanadium ions, especially the V(II) and V(III) which are relatively smaller in molecular size, will diffuse to the other side due to the concentration gradient, and consequently cause self-discharge and capacity fading. Many scientists have attempted various methods to reduce the crossover effect, but they are either expensive or time-consuming, making these approaches difficult to put into practice.

In our current work, we proposed a facile and cost-effective approach using the air-controlled electrospay process to create a thin and uniform conductive coating of different carbon materials onto the surface of the separator. During the coating process, an electric field is applied to force conductive ink particles to move towards the separator substrate, while high-speed but controlled air flow is added in order to help dry the surface and better control the deposition of conductive carbon on the separator surface. Our electrochemical and single cell testing revealed that depositing a very small amount of conductive carbon (< 0.03 mg/cm^2) can drastically change the surface properties of the separator such as the electrical conductivity and permeability, leading to suppressed crossover of vanadium ions, while enhancing the capacity for 28% and energy efficiency for 10% at a current density of 50 mA/cm^2, demonstrating that a facile conductive carbon coating on cost-effective separators has a potential to replace the costly Nafion separator in flow battery systems.

300g Localized High Concentration Electrolytes for Dual Ion Lithium Batteries

Youngbi Kim, Jeong Woo Han

An electrolyte with high Li-metal Coulombic efficiency (CE) and high anodic stability on cathodes is crucial for enabling next-generation high-energy-density lithium (Li)-metal batteries (LMBs). The high concentration electrolyte (HCE) in the current dual-ion battery (DIB) electrolyte achieves a greater capacity than the low concentration electrolyte (LCE), and stable operation at high voltage is made possible by increased electrochemical stability. It also features a solvation structure with a lot of contact ion pairs (CIPs) or cation-anion aggregates (AGG). Nevertheless, high-concentration electrolytes have high viscosity and limited ionic conductivity, which degrades the high-output characteristic of conventional DIBs. In this study, we used an electrolyte type called localized high concentration electrolytes (LHCEs). This decreases viscosity via diluent and stabilizes the contact between the lithium metal and the graphite anode, resulting in steady functioning even at high output. The differences between the three systems, HCE, LCE, and LHCE, were examined using molecular dynamic simulations (MD) and density functional theory (DFT) calculations. Each system was composed of varying proportions of dimethyl carbonate (DMC) and lithium bis(trifluoromethanesulfonyl)imide (LiTFSI), with 1,1,2,2-tetrafluoroethyl -2,2,3,3-tetrafluoropropyl ether (TTE) serving as a diluent for LHCE. To determine how diluents affect the intercalation and desorption of anions, we modelled the solvation structures of anions and cations. As a result, it was found that DMC tightly coordinated with all LiTFSI salt molecules, but not TTE. The radial distribution functions of the Li-O_{DMC} and Li-O_{TTE} pairings were also built using MD simulations over the previous 200 ps. In all three systems under examination, a distinct peak for the Li-O_{DMC} pair was discovered around 2.05 Å, which is consistent with all LiTFSI salts. Being surrounded by DMC solvent molecules shows that DMC is the initial shell of coordination. Moreover, compared to HCE, which has significant coordination with TFSI, LHCE demonstrated significant cooperation with DMC. Coordination of DMC and TFSI yielded similar values in the case of LCE. This demonstrates that TFSI anion mobility is greatest in LHCE. Moreover, DFT simulations of the solvation structure and intercalation behavior of anions were carried out. Comparing the experimental voltage and the calculated value revealed the presence of the same pattern. In order to observe the differences based on the type of anion and the type of solvent, lithium bisfluorosulfonylimide (LiFSI) and ethyl methyl sulfone (EMS) were also used. Hence, LHCE had the highest FSI anion mobility. Surprisingly, radial distribution functions' peak values were lower than when TFSI and DMC were used. These findings will help to improve existing DIB challenges by suggesting new forms of LHCE electrolytes.

300h Engineering Block Copolymer-Derived Porous Carbon Nanofibers As Anodes for Sodium-Ion Batteries

Kenneth Gordon, Cameron Romero, Zhi Liu, Karius Joseph, Emily Broussard, Griffin Usie, Ling Fei

For decades, lithium-ion batteries have seen widespread usage in a variety of applications from consumer electronics such as smartphones to the more recent developments of electric vehicles. As the usage of lithium for batteries increases, one growing concern is the finite amount that could be found within the Earth's crust. An alternative that is being considered is sodium-ion batteries due to how plentiful sodium is in the Earth's crust and similar chemical properties to lithium. One of the main challenges of sodium-ion batteries is the development of anode materials because sodium does not intercalate well with graphite, which is one of the most used materials in lithium-ion batteries. One promising development for anode materials is the use of hard carbon derived from the pyrolysis of polymers such as polyacrylonitrile (PAN) which can have various morphologies and potential for doping. Our work will focus on carbon nanofibers which show high 1-dimensional transport capabilities. With the addition of a sacrificial material such as polymethylmethacrylate (PMMA) that will be removed entirely during the pyrolysis step, PAN/PMMA can be made into porous carbon nanofibers to further increase ion transport. When a PAN/PMMA blend is used, the pores are very random with a high degree of variation of pore characteristics, however studies have shown that the creation of a PAN-b-PMMA block copolymer would allow for more uniform pore structures. Our work intends to study the effects of pore structures on the electrochemical characteristics of porous carbon nanofiber anodes for sodium-ion batteries. The cycling stability data of coin cells made using hard carbon nanofibers derived from electrospun PAN (Nonporous), PAN/PMMA (Random Porous), and PAN-b-PMMA (Uniform Porous), shows their respective discharging specific capacities (mAh/g) at 0.1 A/g average at about 80, 116, and 175. A slight increase can be seen with the addition of random, nonuniform pores to the PAN hard carbon, however, with the highly uniform and controlled pores resulting from the pyrolysis of the PAN-b-PMMA block copolymer, a very significant increase in the discharging specific capacity can be seen.

159a Machine Learning Technique for Core Materials Design Toward Active Systems in Energy Storage and Conversion Reactions

Byungchan Han, Hoje Chun, Minjoon Hong

Nanoparticles are core materials for various electrochemical reactions for energy storage and conversion. Since last several decades the overall systematic efficiency utilizing the reactions have not reached to the level of our target due to several unsolved issues caused by nano-materials' inert functionality, high cost and instability over long-term operation.

In atomic-level computational electrochemistry a new research paradigm for desired materials design has been established through consistent and tight combination of IT-based artificial intelligence (AI) technology and machine learning algorithm standing on supercomputer architectures. This presentation demonstrates the computational methodologies of high-throughput screening of promising nanoparticle candidates for electrochemical energy conversion and storage systems. Using the first-principles density functional theory calculations and frontier realtime and atomic level experimental measurements we acquire reliable and accurate materials properties as input to activate machine learning model. To elevate the accuracy even to higher level we incorporate active-learning and multi-fidelity methods.

It is shown that an AI-based neural-network model is very useful for identifying multi-component electrocatalysts toward three reactions (HER, OER, ORR) at the same time. A computational platform making the pipeline automatic is demonstrated.

159b Modeling and Design of Single-Atom Alloy Catalysts for CO_2 Hydrogenation Reactions

Mie Andersen, Raffaele Cheula

Recently developed machine learning (ML) methods hold great promise for simultaneously reducing the computational cost and increasing the accuracy in catalysis modeling [1]. In this contribution we will discuss recent work aimed at predicting adsorption configurations and energies of complex molecules at the surfaces of transition metals and alloys using a graph-based Gaussian Process Regression model (WWL-GPR [2]). We apply the methodology to study CO_2 hydrogenation (reverse water-gas shift) over single-atom alloy (SAA) catalysts, i.e., diluted bimetallic materials made of a single metal atom dispersed within the surface layer of another metal. SAAs have recently attracted considerable interest since adsorption at the sites offered by their surfaces can break the scaling relationships between adsorption energies and reaction barriers that limit conventional catalysts [3]. We target a wide combinatorial space of elements of the periodic table, which makes a direct study with density-functional theory (DFT) computationally prohibitive. Hence, we produce a database of DFT-calculated energies on a limited number of materials (metals and SAAs) and apply ML techniques for the extrapolation to a much wider range of materials. We include in the study multiple crystal facets that catalyst nanoparticles can expose under reaction conditions. We use the WWL-GPR model to calculate adsorption energies and simpler models (e.g., Brønsted-Evans-Polanyi relations and gradient boosting models with feature vector input) to estimate the activation energies of the new materials. We employ mean-field microkinetic modeling to simulate the reaction kinetics, accounting for the contribution of the different active sites of the catalysts. Then, we apply sensitivity analysis and uncertainty quantification to identify the parameters that can improve the model predictions, and we refine them with additional DFT calculations. The application of the framework to CO_2 hydrogenation reactions (e.g., reverse water-gas shift) allows us to rationalize how reaction mechanisms and catalytic performances (i.e., activity and selectivity) change with the catalyst composition, paving the way toward the design and nano-engineering of SAA catalysts

References

[1] M. Andersen and K. Reuter, *Acc. Chem. Res.* **54**, 2741 (2021).

[2] W. Xu, K. Reuter, and M. Andersen, *Nat. Comp. Sci.* **2**, 443 (2022).

[3] RT. Hannagan *et al.*, *Chem. Rev.* **120**, 12044 (2020).

159c Machine-Learning Driven Exploration of Catalytic Reaction Network

Hyunwook Jung, Sina Stocker, Karsten Reuter, Johannes T. Margraf

The adsorption energy and reaction barrier are two primary inputs for microkinetic modeling of catatlytic processes. However, when involved reaction intermediates are relatively large and diverse, e.g., in syngas conversion, different types of complexity are associated with attaining these parameters. For medium sized molecule, a multitude of possible binding motif often leads to complex potential energy surfaces (PES) with many competing adsorption conformation. This requires the use of global optimization algorithms, which entails computational costs that are prohibitive at the DFT level. Similarly, sophisticated free energy barrier simulations using, e.g., umbrella sampling, are often replaced by the less rigorous harmonic approximation for computational reasons. To tackle these issues, we present a protocol that generates surrogate machine learning potentials for surface chemistry *on-the-fly*. This is applied to both global optimization and free energy calculations pursuing minimal human intervention and computational cost by iteratively updating the training set with the configurations explored by the algorithm.

159d Generative AI for Materials Modeling

Fei Zhou

All data are noisy. This is especially true in condensed matter physics, chemistry, and materials science, as Richard Feynman famously put it, 'Everything that living things do can be understood in terms of the jiggling and wiggling of atoms'. The main challenge to model big scientific data is that they form high dimensional distributions that are intractable. Inspired by recent advancement in generative AI, we propose to adopt probabilistic approaches to high dimensional probability distributions that are tractable and learnable through statistical learning procedures. The efficacy of our approach is demonstrated through a few distinctive case studies, including a denoiser that identifies both ordered/crystalline and disordered/defect atomic structures with state-of-the-art accuracy, and a generative model for molecular and crystal structures. Finally, we propose score dynamics, a general data-driven probabilistic framework to accelerate both fine-level atomic and coarse-level mesoscale simulations of materials, and showcase its applications in barrier-crossing reactions of dihedral angle dynamics in organic molecules, as well as nucleation and growth reactions of liquids.

159e Accelerating the Discovery of Organic Electrode Materials for Li-Ion Batteries: An Integrated Approach Using DFT, Machine Learning, and High-Throughput Virtual Screening

Omar Allam, Seung Soon Jang

Density functional theory (DFT) and machine learning (ML) are employed to investigate the structure-electrochemical performance relationships of organic electrode materials for Li-ion batteries. DFT is employed to compute the redox potential of various organic moieties, providing valuable insights into their electrochemical properties. However, due to the significant computational time required for high-efficacy DFT modeling, it is not ideal for vast material screening. To overcome this limitation, we implement ML techniques, including artificial neural networks (ANN), kernel ridge regression (KRR), and gradient-boosting regression (GBR), for accelerated discovery and assessment of structure-electrochemical relationships. To further enhance the material screening process, we develop a high-throughput virtual screening (HTVS) pipeline consisting of several surrogate learning models with increasing fidelity. The pipeline effectively prunes material candidates at each stage, ensuring only a subset with higher probability for desirable redox potential progresses to more computationally expensive models. By combining DFT, ML, and HTVS, we create an advanced scheme for accurate prediction and analysis of electrochemical activity, facilitating the discovery of novel organic electrode materials with enhanced electrochemical properties. This integrated approach enables more efficient exploration of the materials space and helps to improve the overall performance of the learning model, paving the way for the development of better energy storage devices.

159f Active Learning Exploration of Single Atom Active Sites for Oxygen Reduction and Evolution Reactions

Hoje Chun, Rafael Gomez-Bombarelli, Byungchan Han

Rational design of a single atom catalyst for target catalytic reaction is of great interest. Here, we present a high-throughput screening of local environments of M-N-C (M: 3d transition metals, N: boron, carbon, nitrogen, oxygen, and sulfur, and C: carbon) for oxygen reduction and evolution reaction. We explore the active metal sites with different ligand field interactions inducing the charge and spin states change. We apply the active learning for efficient data generation and equivariant graph convolutional network for surrogate model construction. Our results indicate that the local atomic environments of a single atom catalyst and the spin states can effectively tune the adsorption strength of reaction intermediates. Site-property of active sites electronic structure parameters *i.e.*, atomic charge, coordinated ligands, spin state, *d*-band center energy, etc. Our strategy suggests an effective guidance of the atomic level engineering to activate the center metal for the highly active nonprecious metal single atom catalysts.

141a Stepwise Bulk Process for Enhanced Productivity and Stability of mRNA Vaccines

Minsub Chung, Sangmin Lee, Hyun Jin Kim

The current manufacturing process for mRNA vaccines utilizing the microfluidic technique offers advantages in regulating the size and homogeneity of the lipid nanoparticles (LNP) that form the core components of mRNA vaccine. However, for large-scale production, the capacity of the microchannel-based approach is limited due to difficult scale-up. Additionally, microchannels are susceptible to sedimentation and blockages, necessitating periodic replacement.

We developed a novel stepwise bulk method (SBM) for the manufacturing of mRNA vaccines that is simple, robust, and easily scalable. The new process comprises a series of reactors, each designed to perform specific functions such as mRNA binding and encapsulation. Due to its resistance to clogging and ease of control, this manufacturing process is expected to offer significantly improved productivity. We investigated LNP formation according to the EtOH volume fraction, lipid composition, and N/P ratio to verify the process. Through measurements of size, PDI, encapsulation efficiency, and luciferase luminescence, we confirmed that our method achieves comparable performance to Pfizer's mRNA vaccine produced by the microfluidic method. Finally, we verified that the properties of LNP are maintained as the production scale increases. This new process is expected to contribute to the control of infectious diseases and the development of LNP-based cancer vaccines and therapeutics.

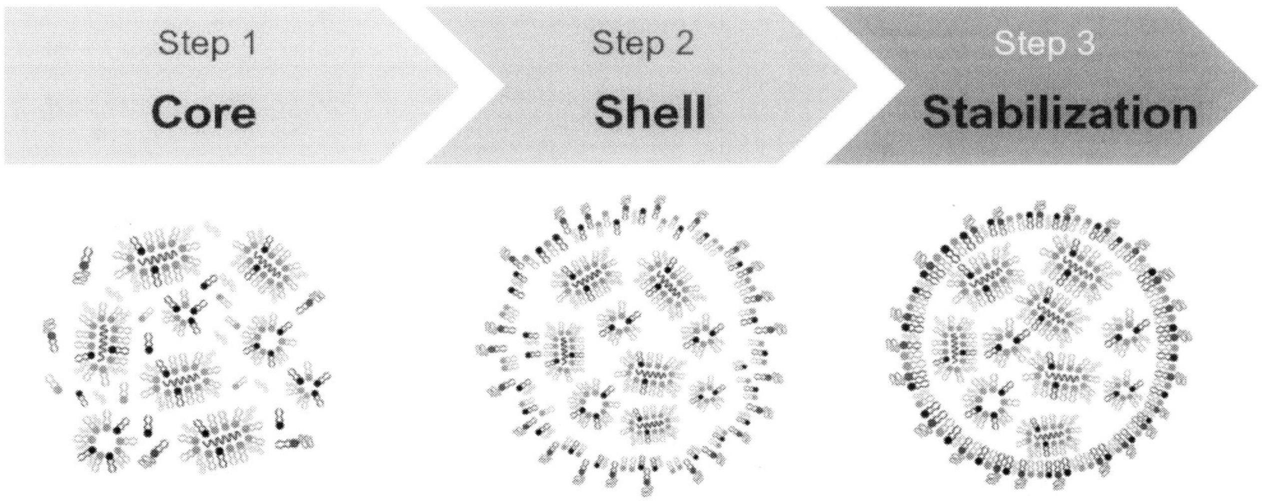
Step 1
Core
Step 2
Shell
Step 3
Stabilization

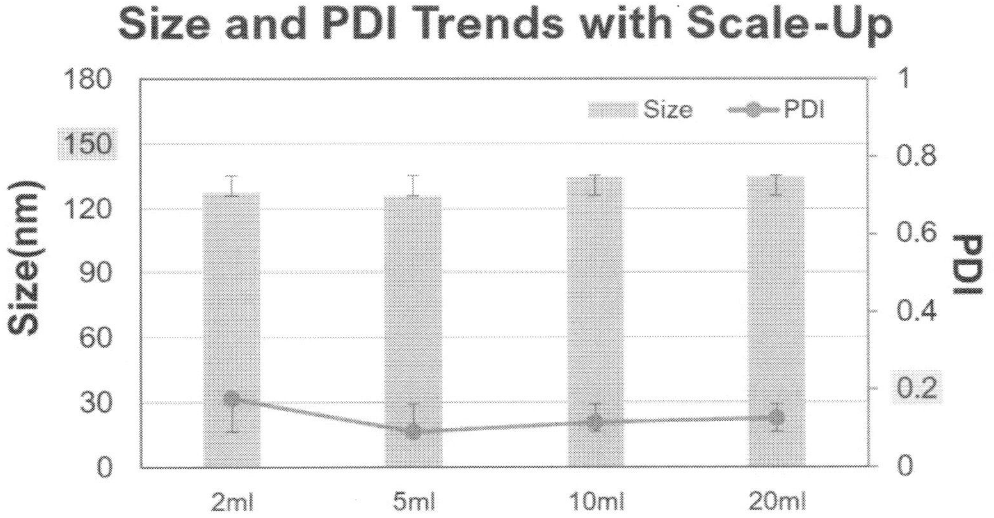
Size and PDI Trends with Scale-Up
Size(nm)
PDI
Size
PDI
180
150
120
90
60
30
0
1
0.8
0.6
0.4
0.2
0
2ml
5ml
10ml
20ml

141b Lipid Nanoparticle Permeabilization and Fusion Quantified By Two High Throughput Fluorescence Assays

Alison Lui, Mariah Arral, Elizabeth Voke, Kathryn Whitehead, Kranthi K. Mandadapu, Markita Landry

Lipid nanoparticles (LNP) containing mRNA have shown success as drug delivery tools and as vaccines against infectious diseases. LNP synthesis is cost-effective and amenable to scaling, however, the LNP design space is high dimensional and it is not well known *a priori* how a given LNP will perform in animal or cell systems as a function of its chemical formulation. Specifically, varying the relative amount and identity of just the helper lipid and ionizable lipid have both been shown to change the efficacy of LNP delivery, cellular toxicity, and organ specificity. In this study, we develop two complementary fluorescence-based assays to quantify the degree and rate of LNP induced permeabilization as well as membrane fusion with a liposomal artificial cell target. By varying nanoparticle parameters such as LNP composition and concentration as well as environmental parameters such as temperature and pH, we probe the dependency of the cellular environment on LNP interactions.

In our work, we find that the ionizable lipid identity, helper lipid identity, target liposomal membrane composition, temperature, and LNP concentration are all controllers of cell membrane permeability. We next explore LNP activity in neutral and acidic environments to determine the effect of different ionizable lipid groups in membrane fusion. We find that the rate and extent of LNP membrane fusion is pH-sensitive, and dependent on the same variables as in the permeability study. For both the permeability and fusion assays, we are able to fit a phenomenological kinetic model to high accuracy to quantify and compare the dependencies on LNP composition, liposome composition, and environmental variables.

By collecting quantitative LNP metrics determined in vitro and relating these to *in vivo* results, we can better understand and inform design of targeted delivery agents for improved nanoparticle therapeutics.

141c Nanoparticle Delivery Doubles the Antinociception Effect of a CGRP Receptor Antagonist in Oral Cancer Pain Models

Parker Lewis, Tu Huu Nguyen, Varun Chokshi, Rachel Pollard, Nigel Bunnett, Brian Schmidt, Nathalie Pinkerton

Problem:

Affecting 30,000 Americans annually, oral cancers are difficult to treat with a modest 50% 5-year survival rate[1]. Oral cancers are extremely painful with increasing pain as the disease progresses[2]. It is well established that the majority of cancer pain is initiated at the tumor site[3]. Pressingly, the therapeutic options to treat cancer pain are limited to opioids, which have profound drawbacks, including sedation, respiratory depression, nausea, constipation, tolerance and dependence[4, 5]. As an alternative to opioids, we target the calcitonin gene-related peptide (CGRP) receptors on cancer cells, which are implicated in cancer pain signaling. Due to the leaky vasculature and enhanced lymphatic drainage characteristic of oral cancers, maintaining drug concentrations in the tumor microenvironment (TME) is challenging[6]. This limits the duration of efficacy of free small molecule CGRP antagonists. **Our goal** is to enhance the duration and intensity of analgesia by using CGRP antagonist loaded nanoparticles, which (1) are better retained in the TME due to their large size (100 nm) compared to small molecule drugs (10 Å) and (2) can be designed for sustained drug release.[7]

Methods:

The CGRP antagonists olcegepant was encapsulated into PLA-PEG nanoparticles using the Flash NanoPrecipitation (FNP) process, in which nanoparticles are assembled via diffusion limited self-assembly driven by hydrophobic interactions. The hydrophobic ion pairing (HIP) nanoformulation approach was used to control the olcegepant release rate. Nanoparticles were characterized by standard physicochemical characterization techniques. To test antinociception in vivo, a human squamous cell carcinoma (HSC-3) cancer xenograft paw and tongue model were used. After the mice developed cancer nociception, the mice were injected with either with free olcegepant (500 nM) or olcegepant-loaded nanoparticle (500 nM). Saline and empty NPs served as controls. To measure cancer nociception development in paw or tongue and to determine the treatment efficacy, we used the von Frey filament and Hargreaves assays for mechanical and thermal allodynia, respectively.

Results:

Olcegepant was successfully encapsulated into 100 nm PLA-PEG nanoparticles using the HIP nanoformulation approach with pamoic acid as the counter ion. Nanoparticles had an olcegepant drug loading of 5 wt%. The olcegepant FNP encapsulation efficiency was 60%. The nanoparticles exhibited a slow drug release rate with 80% drug release after 24 hours. In the cancer paw model, free olcegepant reduced 75% mechanical and 50% thermal cancer nociception for 6 hours post-administration. The olcegepant-loaded NPs-loaded reduced 98% mechanical and 100% thermal cancer nociception for 12 hours post-administration with a persistent reduction in pain lasting up to 24 hours.

Implications:

Inhibiting CGRP receptor signaling is a promising approach for the treatment of cancer pain. By using a nanoparticle drug delivery approach to deliver olcegepant in the TME, the duration and intensity of antinociception can be dramatically increased. This opens up a new avenue beyond opioids for cancer pain therapeutics.

References:

1. Sciubba, J. J., American Journal of Clinical Dermatology 2001, 2 (4), 239-251.
2. van den Beuken-van Everdingen, M. H et al.,. Ann Oncol 2007, 18 (9), 1437-49.
3. Kolokythas, A.; et al., J Pain 2007, 8 (12), 950-3.
4. Bruera, E.; Kim, H. N., Jama 2003, 290 (18), 2476-2479.
5. Mercadante, S., Cancer 1999, 86 (9), 1856-1866.
6. Sato, S., et al., JCI Insight 2019, 4 (23).
7. Ashton, S., et al., Science Translational Medicine 2016, 8 (325), 325ra17.

141d Block-Copolymer Stabilized Lipid Nanoparticles: A Novel Platform for Targeted Gene Delivery

Sai Nikhil Subraveti, Bumjun Kim, Robert K. Prud'homme

Lipid nanoparticles (LNPs) have demonstrated their enormous potential as delivery vehicles, as evidenced by the approval of two COVID-19 mRNA vaccines. Gene delivery is the next frontier for LNPs. The current state-of-the-art LNPs include four main components: an ionizable lipid, a helper lipid, cholesterol, and a PEG-lipid. The PEG-lipid, while only at a small mole fraction, plays a vital role in LNP stability, transfection efficiency, and liver targeting. Indeed, the liver targeting of LNPs is due to the rapid shedding of PEG-lipid upon I.V. administration, followed by the subsequent adsorption of apolipoprotein E (ApoE), leading to low-density lipoprotein receptor (LDLR)-mediated uptake by hepatocytes. Although the PEG-lipid desorption is instrumental in exploiting the well-established delivery route to the liver, it severely hinders the ability of LNPs to target non-hepatic cells passively or actively. Therefore, to reduce liver targeting and to realize non-hepatic cell targeting, a novel LNP platform with a tightly anchored stabilizer is crucial.

We present LNPs prepared using block copolymer (BCP) stabilizers that do not shed, but rather stay anchored on the LNP surface. We systematically optimized several parameters (composition, concentration, payload) and identified the best BCP-stabilized LNP formulations with desired properties such as long-term stability, optimal sizes (~100 nm), high encapsulation efficiencies (>90%), and high transfection efficiencies in different immortal cell lines. While permanent PEG stabilizing layers on LNP surfaces might be expected to compromise either cellular internalization or endosomal escape. Surprisingly, these block copolymer PEG LNPs transfected more effectively than the corresponding PEG-lipid stabilized LNPs. Hence, this new construct enables the targeting of LNPs in a way that is not possible with PEG-lipids. Likewise, these BCPs can be conjugated with large ligands (antibodies) to target non-hepatic cells. The capability to conjugate targeting ligands and antibodies to block copolymer LNPs, without compromising transfection, potentially enables gene delivery formulations which cannot be delivered using conventional LNP constructs.

141e Rational Design of Silica-Based Nanoparticles for Overcoming Barriers of siRNA Delivery in Relapsed Prostate Cancer Applications

Angelea Maestas-Olguin

The use of exogenous siRNA technology to modulate aberrant protein expression resulting from genetic mutations is a promising therapeutic approach for treatment of diseases such as prostate cancer (PC). The promise of siRNA-based therapeutics is dependent on the development of platforms that effectively protect siRNA from nuclease degradation and deliver the siRNA to the cytosol of target cells. The establishment of nanoparticle-based siRNA delivery platforms have been investigated to mitigate the cost and safety issues associated with viral delivery platforms. In this work, we present the development and characterization of a lipid coated mesoporous silica nanoparticle (LC-MSN) to address the need for safe and effective siRNA delivery across various biological barriers in the relapsed prostate cancer landscape. The LC-MSN utilizes calcium silicate nanogating over the silica core to efficiently load (80%) and release (>80%) siRNA in relevant media, as analyzed by tracking a fluorescently tagged siRNA cargo. Additionally, the calcium silicate core is encapsulated in a lipid bilayer aimed to improve the LC-MSN biocompatibility and enables the integration of GRP78 minibodies for targeted delivery to PC cells. The cytotoxicity, uptake and intracellular fate of the LC-MSN is investigated within our work using the LNCaP PC cell line. Furthermore, our work utilizes an *ex ovo chick* chorioallantoic membrane model (CAM) to assess LC-MSN system vascular margination, binding, circulation time and stability. The structural and chemical versatility of the silica nanoparticle core along with a biocompatible lipid coating makes the LC-MSN a promising candidate for siRNA delivery within the relapsed PC landscape.

141f Influence of Protein Corona on Lipid Nanoparticle Uptake

Elizabeth Voke, Mariah Arral, Rebecca L. Pinals, Henry Squire, Alison Lui, Anthony Lavarone, Kathryn Whitehead, Markita Landry

Lipid nanoparticles (LNPs) are the most clinically advanced nonviral RNA-delivery vehicles, demonstrating successful delivery of mRNA-based vaccines such as the Moderna and Pfizer/BioNTech SARS-CoV-2 vaccines. Previous work to improve LNP potency and targeting capabilities has primarily focused on variations in LNP composition. Although this approach has led to some efficient LNP-mediated gene expression and delivery to tissues outside of the liver, we still do not understand the fundamental mechanisms that drive increased potency or organ- and cell-type-specific biodistribution. As these LNPs travel through biological tissues and compartments, biomolecules such as proteins spontaneously interact with the nanoparticles, forming an associated protein corona. Protein recruitment to the LNPs has been shown to play a role in enhanced delivery, yet protein corona formation on such soft, biomimetic nanoparticles remains poorly understood and is largely unexplored due to difficulties associated with isolating the adsorbed protein fraction. In this work, we have developed a quantitative, label-free mass spectrometry-based proteomics approach to probe the nano-bio interface of a novel class of LNPs for mRNA delivery. We have optimized a separation method that, for the first time, limits the contamination of endogenous biofluid nanoparticles in protein corona isolates from LNPs. With this method, we discovered proteins with a high affinity for the LNPs such as ApoE and C-reactive protein, in agreement with the minimal previous literature on such lipid-based carriers. Our data suggests that this method is a promising avenue for corona characterization. We apply our workflow to investigate how altering the LNP formulation – variations in helper lipid within the lipid nanoparticle – impacts the formation of the protein corona in various relevant delivery bioenvironments and explore the impact of these protein targets on expression of mRNA in HepG2 human liver cells. By understanding these fundamental protein-nanoparticle interactions, we aim to rationally tune the design of these mRNA-based biotechnologies for improved translation to clinical practice.

141g Reactive Oxygen Species (ROS)-Responsive Doxorubicin Prodrug for Breast Cancer Treatment

Mina Jafari, Sophia Schutte, Lisa Privette Vinnedge, Joo-Youp Lee

Toxic side effects are common with systemic chemotherapies and impair the successful completion of treatment for many patients with cancer. Doxorubicin (Dox) is commonly used to treat various cancers including breast, bladder, lymphoma, and leukemia, but ~4-36% of patients experience potentially lethal cardiotoxicity in a dose-dependent manner. Although nanoparticle (NP)-based therapeutics have made tremendous progress, it is still a great challenge to meet the goals due to the two desired but conflicting concepts of sustained and controlled release. A stimulus-responsive linker approach can address these challenges by deactivating potent but toxic drugs upon conjugation to a polymer and reactivating them upon release triggered by a stimulus-responsive mechanism. Reactive oxygen species (ROS) can be used as a powerful internal stimulus, as the ROS level in cancerous cells (e.g., ~100 mM) can go up to ~5,000 times higher than those in healthy cells (e.g., ~20 nM).

Here, we present two ROS-responsive Dox prodrug-based nanoparticle (NP) systems that can deliver Dox in a sustained manner during systemic circulation while achieving controlled drug release at tumor sites. Our ROS-responsive Dox prodrug was physically encapsulated or chemically conjugated. The NP designs showed average sizes of <~100 nm with very stable Dox release profile in PBS at pH 7.4. While both NP designs showed ROS-responsive Dox releases up to ~70-80% within 48 hrs under 50-100 mM H_2O_2 and horseradish peroxidase (HRP) concentrations, the physically-encapsulated NPs showed faster Dox release in the first few hrs. The surface modification of NPs with hyaluronic acid (HA), a natural polysaccharide that can specifically bind to CD44, resulted in an increase in cellular uptake and efficacy against MDA-MB-231 breast cancer cell line. The therapeutic efficacy and off-target toxicity of our NPs were also investigated *in vivo* using an orthotopic xenograft tumor model using the MDA-MB-231 cell line.

185a NSEF Young Investigator Award Winner: Nanoscale science and Engineering of Atomically Thin Membranes for Next-generation Healthcare, Energy, & Environmental Applications

Piran Kidambi

Atomically thin 2D materials offer fundamentally new opportunities to study, understand and control mass transport at the sub-nanometer-scale. Specifically, they allow for quantum tunneling and size-selective ionic/molecular sieving. I will discuss our recent work in 2D material synthesis and processing to enable i) large-area atomically thin Helium barriers, ii) fully functional nanoporous atomically thin membranes for desalination, dialysis and molecular separations, iii) in-situ bottom-up formation of precise nanopores in 2D materials, iv) development of facile methods to probe sub-nanometer scale defects in centimeter scale 2D-single-crystals, v) roll-to-roll manufacturing of atomically thin membranes, vi) novel size selective defect sealing for functional large-area 2D membranes and vii) sub-atomic species transport through atomically thin membranes for advancing energy conversion/storage applications. Finally, I will discuss some of our efforts to move these technologies from the lab space to the commercial arena.

185b NSEF Forum Award Winner: Making optical materials from metal oxide nanocrystals

Delia Milliron

As optical materials, metal oxides usually fulfill basic functions, like transparent window glass or particles that scatter light in bright white paint. We make materials with more sophisticated optical functionality, with highly tunable and even dynamic spectral response, from metal oxides by doping them with deliberate impurities and by shrinking their dimensions to the nanoscale. Replacing a few percent of the metal ions in a wide band gap semiconductor like indium oxide with aliovalent dopants like tin results in the unusual combination of optical transparency with metallic conductivity in indium tin oxide (ITO). When synthesized in colloidal nanocrystal form, ITO maintains its visible transparency, while the confinement of free electrons results in the strong absorption of infrared light and the concentration of energy into nanoscale volumes around the nanocrystals due to localized surface plasmon resonance (LSPR). Based on their LSPR characteristics, ITO nanocrystals can boost sensitivity of molecular detectors or act as remote monitors of biochemical electron transfer reactions. The infrared absorption of ITO and other metal oxide nanocrystals is strongly modulated by electron transfer, enabling fabrication of smart windows that dynamically control solar heat gain and visible glare. Based on their optical tunability and dynamic responsivity, these nanocrystals have the potential to make solar cells and buildings more efficient, drive selective chemical reactions, harness solar energy, and more.

299a Invited Talk: Carbon Nanotube Quantum Well Defect Emission for Machine Learning-Guided Diagnostics

Daniel Heller, Mijin Kim, Anand Jagota, YuHuang Wang, Ming Zheng

We employ the photoluminescence of single-walled carbon nanotube (SWCNTs), and covalent sp3 quantum well defects on SWCNTs, to develop new diagnostic methods for cancer and other diseases. Serum biomarker measurements are widely used for diagnosis, but these markers largely provide low sensitivity and specificity. We developed a method using defect-modified SWCNTs to identify a "disease fingerprint" through the collection of large data sets of molecular binding interactions to an array of quantum defect-modified carbon nanotubes. We found that a library of modified SWCNTs exhibited differentiated spectral variation in response to an ensemble of molecular binding events in patient serum. Via machine learning algorithms, we built a prediction model of nanosensor responses that reliably identified ovarian cancer substantially better than the established, FDA-approved biomarker, CA125. We have expanded this approach to other indications without known biomarkers, providing a general method to identify diseases.

299b Hardware and Sensor Optimization of a Nanotechnology Enabled Implantable Sensor for Hormones in Marine Animals

Volodymyr Koman, Xiaojia Jin, Manki Son, Michael Strano

Biologging has yielded tremendous insights in ecological biology by enabling the collection of data in animals free-roaming in their environments. However, measurements have largely been limited to temperature, pressure, and movement. Physiological data of animals' biomarkers would create valuable orthogonal datasets and produce more nuanced understanding of organisms in the context of their environments and behaviors. Nevertheless, successes in collecting such biochemical information remain absent. In this work, we explore hardware and sensor optimization of nanotechnology-enabled implantable sensor for tracking hormone levels in marine animals. We have extensively studied the performance of nanosensors based on polymer-wrapped single-wall carbon nanotubes (SWNT) embedded within a biocompatible poly(ethylene glycol) diacrylate (PEGDA) hydrogel under various temperatures, illumination conditions, and sensor concentrations. We have further prototyped a miniaturized fluorescent system integrated into commercially available acoustic tags used in marine biology. An integrated nanosensor hydrogel reached 100 nM limit of progesterone detection as an example of important hormone in marine animals. The developed form-factor will complement the presently collected data by providing insights into the physiological state of the animals in the context of their behavior and environments. With these new data, biologging studies will reach a new level of sophistication and nuanced understanding.

299c Covalent Attachment of Horseradish Peroxidase to Single-Walled Carbon Nanotubes for Hydrogen Peroxide Detection

Francis Ledesma, Shoichi Nishitani, Francis J. Cunningham, Linda Chio, Aishwarya Murali, Markita Landry

Hydrogen peroxide (H_2O_2), a reactive oxygen species (ROS), is critical to the regulation of mammalian cellular processes such as apoptosis and response to oxidative stress. However, unregulated cellular levels of H_2O_2 leads to erratic DNA mutation and damage, implicating it as a biomarker for diseases such as cancer, Parkinson's, Alzheimer's, and Huntington's. It is thus important to develop sensitive and selective technologies to sense elevated cellular H_2O_2 levels for improved disease diagnostics. Current techniques for H_2O_2 sensing involve chemiluminescent or fluorescent dyes, but these suffer from instability in dynamic environments, photobleaching, and low spatiotemporal resolution.

Here, we developed a nanosensor for detection of H_2O_2. We base our nanosensor on single-walled carbon nanotubes (SWCNTs), owing to their intrinsic near-infrared (nIR) fluorescence, high aspect ratio, and photostability. The selectivity of our nanosensor is conferred by covalently attaching horseradish peroxidase (HRP) to the SWCNT surface, a key enzyme in the oxidative stress response pathway of the horseradish plant. We employed covalent surface modification of SWCNTs with cyanuric chloride, maintaining intrinsic nIR fluorescence[1] while priming the SWCNT surface for further modification with common protein conjugation handles such as carboxylic acid, primary amine, and thiol groups.[2] We functionalized HRP with maleimide groups via NHS-lysine conjugation with a linker named Sulfo-SMCC. We subsequently incubated maleimide-functionalized HRP with thiol-functionalized SWCNTs to covalently attach HRP to SWCNTs without compromising enzymatic activity.

We characterized the nanosensor's physical and optical properties to confirm successful protein conjugation while maintaining intrinsic SWCNT fluorescence and HRP enzymatic activity. Additionally, we assessed the nanosensor's concentration-dependent response to H_2O_2 in solution and demonstrated a limit of detection of 31 uM. Furthermore, we demonstrated the nanosensor's selectivity for H_2O_2 by its lack of response to an analyte panel of related reactive oxygen species and other biologically relevant analytes. Finally, we immobilized the nanosensor on glass to confirm its fluorescent signal is reversible, which we can use to sense elevated peroxide levels in cell lysate from oxidatively stressed MCF-7 cells. Taken together, these results demonstrate covalent attachment of enzymes to SWCNTs with a chemical strategy that preserves both intrinsic SWCNT fluorescence and enzyme function. We anticipate this platform can be adapted to covalently attach other proteins of interest such as other biologically relevant enzymes for sensing and larger protein cargo for cellular delivery including CRISPR-Cas9.

References

1. Setaro, A. *et al.* Preserving π-conjugation in covalently functionalized carbon nanotubes for optoelectronic applications. *Nat. Commun.* **8**, 14281 (2017).
2. Chio, L., Pinals, R. L., Murali, A., Goh, N. S. & Landry, M. P. Covalent Surface Modification Effects on Single-Walled Carbon Nanotubes for Targeted Sensing and Optical Imaging. *Adv. Funct. Mater.* 1910556 (2020) doi:10.1002/adfm.201910556.

299e High-Throughput Evolution of Near-Infrared Oxytocin Nanosensors

Jaquesta Adams, Nicole Navarro, Esther Leem, Xiaoqi Sun, Tavio Arias, Markita Landry

Neuropeptides act broadly as neurotransmitters, neurohormones, and neuromodulators throughout the body, implicating these macromolecules in a variety of physiological processes and their dysfunction in multiple neurological disorders and neurodegenerative diseases. Neuropeptide imaging can provide insight into their signaling pathways and spatiotemporal dynamics thus connecting their broad activity to both function and pathology. It is of great interest, then, to develop tools with the requisite metrics to enable *in vivo* neuropeptide imaging in the brain. Synthetic nanoparticles offer relatively simple design and synthesis, facile signal readout, and rapid detection response along with orthogonality to biological processes. Single-walled carbon nanotubes (SWCNT) are one-dimensional nanomaterials with intrinsic near-infrared fluorescence, whose surface modification with amphiphilic polymers confers analyte-specific sensing capabilities. Herein, we present on the results of high-throughput identification of DNA-SWCNT conjugates with high oxytocin affinity, and characterize these constructs for *in vitro* response to oxytocin.

299f Glycopolymer-Wrapped Carbon Nanotubes for Sars-Cov-2 Spike Protein Detection

Brandon Heppe, Geyou Ao, Xue-Long Sun

There is a critical need for establishing versatile and effective viral detection platforms since the global pandemic of SARS-CoV-2 and the initial limitations in testing capabilities. Carbohydrate-mediated interactions play a key role in many cellular events, including viral and bacterial infection, that is facilitated by selective interactions of carbohydrates with carbohydrate-binding proteins (CBPs). Glyconanomaterials with specific carbohydrate functionalities are known to mimic the multivalent carbohydrate-protein interactions occurring in biology. Our previous work has investigated the specific interactions of glycopolymer-wrapped single-wall carbon nanotubes (Glyco-SWCNTs) with CBPs using lectins. Specifically, Glyco-SWCNTs were created *via* noncovalent complexation with lactose-containing glycopolymers. Here, we build upon our previous findings to optically detect the specific binding of SARS-CoV-2 spike protein to carbohydrates by utilizing the intrinsic near-infrared fluorescence of Glyco-SWCNTs. Ultimately, this work will lead to the development of microbial nanosensors and further enable the mechanistic understanding of carbohydrate-protein interactions occurring in pathogen infection.

299g Decoding Early Stress Signaling Waves in Plants Using Multiplexed Nanosensors

Mervin Ang, Jolly Saju, Thomas K. Porter, Sarangapani Sreelatha, Duc-Thinh Khong, Suh In Loh, Song Wang, Gajendra Pratap Singh, Nam-Hai Chua, Michael Strano, Rajani Sarojam

Plants are heavily influenced by environmental stresses that can have a significant impact on their growth and productivity. To cope with these stresses, plants activate a complex signaling cascade that leads to stress resistance or adaptation. However, due to the lack of real-time sensing tools capable of simultaneous measurements of multiple signaling molecules, the temporal ordering and composition of stress signaling cascades and their specificity to stresses remain largely unknown. In this work, we developed and validated a novel carbon nanotube-based sensor for salicylic acid (SA) and paired it with a H_2O_2 nanosensor, as these species are two key early stress signaling molecules that interact with each other to mediate stress responses. The sensors were used to monitor stress-induced H_2O_2 and SA signals in *Brassica rapa subsp. Chinensis* (Pak choi) plants subjected to different stress treatments, including light stress, heat stress, pathogen stress, and mechanical wounding. The nanosensors reported distinct dynamics and temporal wave characteristics of SA and H_2O_2 generation in tandem for each type of stress stimuli. For all stresses, the H_2O_2 wave was observed within minutes of stress perception, followed by recovery within the first hour. For SA, however, production onset occurred at different time points within two hours of stress perception, and no SA was measured for mechanical wounding. With insights gained from sensor multiplexing, we constructed a mathematical model that describes a potential mechanism by which stress specificity is captured by the unique H_2O_2 waveform generated shortly after stress perception by the plant. Our results demonstrate that sensor multiplexing is promising for elucidating stress signaling mechanisms in plants, which is a key step toward the development of climate-smart crops and pre-symptomatic diagnoses of stresses in field settings.

299h Imaging Dopamine Modulation in Brain Acute Slices for Cocaine Addiction Study with Non-Genetically Encoded Nanosensors

Nicholas Ouassil, Natsumi Komatsu, Markita Landry

Drug addiction is a pressing issue in modern American society that demands thorough research into its mechanisms. Cocaine locomotor sensitization is a well-established model for understanding the physiological effects of cocaine abuse. Previous studies have established loss of dopamine D2 autoreceptors in striatal tissue as a consequence of locomotor sensitization, yet how these differences in autoreceptor expression affect dopamine signaling remains unclear. Herein, we utilized non-genetically encoded near-infrared fluorescent nanosensors to image dopamine modulation in live acute coronal brain slices. By leveraging dopamine D2 receptor agonists along with the dopamine nanosensor, we developed a method to identify changes in density of functional D2 autoreceptors and dopamine signaling aberrations that accompany addiction, at the level of individual dopamine synapses. We apply the method to compare changes in D2 autoreceptor activity in the Nucleus Accumbens core in cocaine sensitized and saline (control) mice after prolonged cocaine exposure (5 days). Surprisingly, we find no significant changes in D2 autoreceptor activity between cocaine sensitized and saline mice. Furthermore, we confirm dopamine signaling changes from application of D2 Agonists (quinpirole) are from loss of individual dopamine release sites rather than changes in the quantity of dopamine released, or changes in dopamine reuptake kinetics. Our study provides a route to understand the changes in neuromodulation that underlie the neurobiology of addiction and other dopamine-centric neurological conditions.

299i Wide-Range Size Tuning of Conjugated Polymer Nanoparticles for Multi-Color Bioimaging

Noriko Nakamura, Mikito Tokumaru, Seiichi Ohta

1. **Introduction**

Fluorescent nanoparticles have attracted significant attention as strong tools for bioimaging and biosensing. The representative example is quantum dot (QD), which shows size-dependent optical property due to the quantum confinement effect. QDs exhibit high photo-stability and tunable emission wavelength, but on the other hand, their toxicity caused by heavy metal components has been a potential concern. Recently, conjugated polymer nanoparticles, which are called polymer dots (Pdots), have been investigated as a new class of fluorescent nanoparticles without toxic heavy metals [1]. Whereas QDs show photoluminescence only when their size is within the single-nm range, Pdots maintain their optical property with a size up to several hundred nanometers.

The physiological property of nanoparticles, including size, shape, and surface chemistry, substantially affect their pharmacokinetic behavior in the biological systems. Particularly, the correlation between the size of nanoparticles and their interaction with cells and tissues in our body has recently attracted considerable research interest. Considering physiological phenomena, such as hepatic filtration, tissue extravasation, and kidney excretion, the size is a key parameter in the biodistribution and blood half-life of nanoparticles. In addition to enhanced permeability and retention (EPR) effect, tumor affects immunological system in whole body, which may alter the clearance behavior of nanoparticles from the bloodstream and its size dependency.

In general, Pdots are synthesized *via* nanoprecipitation method using commercially available conjugated polymers, which have delocalized π-electron system. The nanoprecipitation method demonstrated Pdot synthesis with a reliable size, usually up to tens of nanometers. Afterwards, many literatures have been investigated Pdot synthesis using stabilizing agents such as polystyrene grafted with ethylene oxide as well as poly(styrene-co-maleic anhydride) (PSMA). These stabilizing agents contribute to the colloidal stability of larger Pdots as well as the enhancement of biocompatibility.

Herein, we propose a size tunable Pdot synthesis strategy for controlling its behavior in the biological systems [2]. In the nanoprecipitation method, PSMA, which is widely used as a stabilizing agent, is initially hydrolyzed into a hydrophilic maleic acid copolymer to stabilize the particle surface. We succeeded in the size tuning of Pdots within the range from 15 nm to 220 nm by controlling the hydrolysis rate of PSMA and particle growth rate, and then demonstrated the ability of size tunable Pdots as a biomedical tool.

2. **Materials and method**

Pdots were synthesized based on the nanoprecipitation method using three different conjugated polymers; poly(9,9-dioctylfluorene-*alt*-benzothiazole) (F8BT), poly[2-methoxy-5-(2'-ethylhyxyloxy)-1,4-phenylene vinylene] (MEH-PPV), and poly(9,9-di-*n*-octylfluorenyl-2,7-diyl) (PFO). Conjugated polymer and PSMA were separately dissolved into tetrahydrofuran (THF), and then the mixture of these two solutions was added to water or 10 mM acetate buffer under sonication and stirring. After Pdots were formed, the THF was removed by evaporation at room temperature, and then purified by ultracentrifugation. The size of synthesized Pdots was evaluated using transmission electron microscopy (TEM) and dynamic light scattering (DLS), and their optical property was characterized by UV-vis spectrometer and fluorescent spectrometer. The surface of Pdots was modified with poly(ethylene glycol) (PEG) in order to increase the biocompatibility. The conjugation of PEG onto Pdots was evaluated by measuring zeta potential and Fourier transform infrared spectroscopy (FT-IR). The performance of PEG modified Pdots with different sizes as a bioimaging probe was then examined using mouse fibroblast (NIH/3T3) cells. The Pdots internalization into cells was evaluated using confocal laser microscopy and flow cytometry.

3. **Results and discussion**

Firstly, size-tunable synthesis method of Pdots was developed using F8BT as a conjugated polymer. Conjugated polymer and PSMA were dissolved in THF, followed by addition to the water phase with varying pH and ionic strength to synthesize Pdots. The synthesized Pdots were of spherical morphology and unimodal size distribution for all conditions. The size of Pdots decreased from approximately 220 nm to 60 nm with increasing pH from 5.4 to 6.2 when synthesized using 10 mM acetate buffer with 80 mM NaCl as the water phase. The decrease in Pdot size would be attributed to the accelerated hydrolysis rate of maleic anhydride units of PSMA in the basic condition. The hydrodynamic size of Pdots determined by DLS was also consistent with the TEM observation, thereby confirming the successful tuning of Pdots size by controlling pH of the water phase. In addition to the size-tuning based on pH, the size of synthesized Pdots also could be controlled by changing the volume of deionized water as the water phase during nanoprecipitation. When the volume of water increased, the size of Pdots decreased from 30 nm to 15 nm. Therefore, we succeeded in tuning the size of Pdots within the range from 15 nm to 220 nm. It is also confirmed that this size-tuning strategy can be adapted to other conjugated polymers, MEH-PPV and PFO.

The optical property of Pdots is crucial for their performance as fluorescent probes for bioimaging. The adsorption peak of synthesized Pdots was slightly red-shifted with increasing particle size, which could be due to the extension of the delocalization of π-electron conjugation. Meanwhile, Pdots showed the same fluorescent emission spectra regardless of their size. Quantum yield also did not change significantly, regardless of the size. For investigating the performance of Pdots with different diameters as fluorescent probes, the surface of Pdots was modified with PEG to minimize non-specific interaction with molecules in the physiological environment such as proteins. The surface zeta potential of Pdots increased after the reaction, which indicated that the PEG modification successfully proceeded. The successful surface modification of Pdots was also suggested by their FT-IR spectroscopic spectra before and after PEGylation.

The performance of Pdots with different sizes as bioimaging probes was then examined using NIH/3T3 cells *in vitro*. PEGylated Pdots were incubated with cells for 4 hours, followed by the observation using confocal laser microscopy. Cellular uptake of Pdots was successfully observed in all sizes within 60 nm to 220 nm. Furthermore, Pdots were colocalized with endosomes and lysosomes stained with LysoTracker, which suggested their internalization into cells through endocytosis. To further investigate their cellular uptake behavior, the fluorescent intensity from different sizes of Pdots internalized into each cell was quantified using flow cytometry. Cells were incubated with the presence of PEGylated Pdots under the same condition described above, and then trypsinized cells were analyzed using a flow cytometer. It was found that the Pdots showed higher fluorescent intensity in each cell with their increasing size. Therefore, it was confirmed that our size-tuned Pdots can demonstrate the size effect of nanoparticles in the biological environment.

4. **Conclusions**

In this study, we demonstrated the size-tunable synthesis of Pdots based on a nanoprecipitation method by regulating the hydrolysis rate of PSMA as a stabilizing agent. We found that the size of Pdots was successfully controlled by pH of the water phase during nanoprecipitation because it determined the hydrolysis rate of maleic anhydride units of PSMA. Additionally, the size of Pdots also tuned within the range smaller than 30 nm by changing the volume of water phase to control the particle growth rate. Taken together, facile on-demand size tuning of Pdots between 15 nm to 220 nm was achieved via optimizing reaction conditions. These strategies work as long as acid anhydrides are used during nanoparticle formation, and are expected to be adapted not only to fluorescent conjugated polymers but also to other hydrophobic polymers. We also succeeded in evaluating functions of size tuned Pdots using cultured cells. It is expected that size tunable Pdots will significantly broaden the application of Pdots as fluorescent materials in the biomedical field.

5. **References**

[1] C. Wu and D. Chiu, *Angew. Chem., Int. Ed. Engl.,* 2013, **52**,3086–3109. [2] N. Nakamura, N. Tanaka and S. Ohta, *RSC Adv.*, 2022, **12**, 11606-11611.

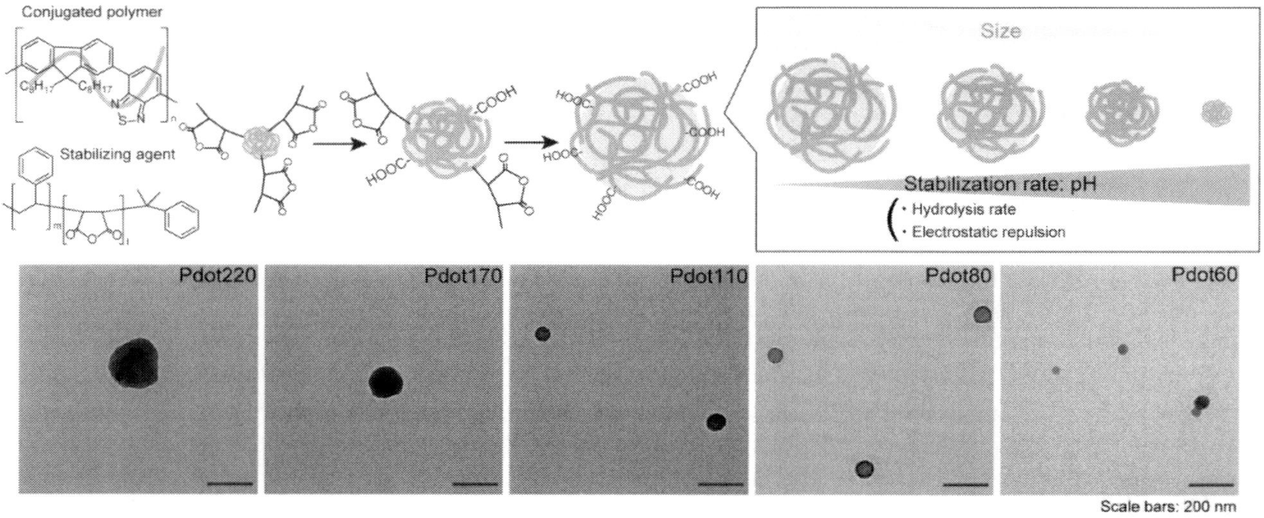

51a Direct Air Capture (DAC) of CO_2 Using Porous Sorbent Materials Coupled with Scalable Processes

Christopher Jones

Worldwide energy demand is projected to grow strongly in the coming decades. Even with unprecedented growth rates in the development of renewable energy technologies such as solar, wind and bioenergy, the world will continue to rely on fossil fuels as the predominant energy source for at least the next decade. Simultaneously, due to decades of inaction, current climate models as well as the recent IPCC AR6 Climate Change Report state that limiting warming to <2°C will require large scale deployment of negative emissions technologies (NETs). NETs, which remove CO_2 from the atmosphere, are projected to be needed at a scale of 10 Gt/y by 2060, yet today, virtually none have been deployed.[1] NETs may be natural or technological, with one of the most scalable technological approaches being the direct capture of CO_2 from the air, or "direct air capture" (DAC).[2,3] Because of the ultra-dilute nature of CO_2 in air, the separation of CO_2 from this mixture presents a significant engineering challenge.

In this lecture, I will describe the design and synthesis, characterization and application of supported amine materials and other nanoporous materials that we have developed as cornerstones of new technologies for the removal of CO_2 from ultra-dilute (air) gas streams.[4] The amine chemisorbents efficiently remove CO_2 from simulated flue gas streams, and the CO_2 capacities are enhanced by the presence of water, unlike the case of physisorbents such as zeolites. We will describe the development of these materials, how they integrate into scalable DAC technologies, as well as their key physicochemical structure-property relationships. DAC technologies offer an interesting case study for the parallel design of materials, unit operations, and processes in chemical engineering. Contemporary challenges in DAC will be discussed. [5]

References

1. https://nas-sites.org/dels/studies/cdr/
2. https://chemistryshorts.org/
3. "Direct Capture of CO_2 from Ambient Air." E. S. Sanz-Pérez, C. R. Murdock, S. A. Didas, C. W. Jones, *Chemical Reviews*, **2016**, 116, 11840-11876.
4. "Amine-Oxide Hybrid Materials for CO_2 Capture from Ambient Air." S. A. Didas, S. Choi, W. Chaikittisilp, C. W. Jones, *Accounts of Chemical Research*, **2015**, 48, 2680-2687.
5. "Research Needs Targeting Direct Air Capture of Carbon Dioxide: Material & Process Performance Characteristics Under Realistic Environmental Conditions." F. Kong, G. Rim, M. Song, C. Rosu, P. Priyadarshini, R. P. Lively, M. J. Realff, C. W. Jones, *Korean Journal of Chemical Engineering* **2022**, 39, 1-19.

51b Thermally Stable Single Atom Catalysts for Energy Conversion

Abhaya Datye

Over the past decade, single atom catalysis has evolved from being an academic curiosity to one of the most widely studied methods for the synthesis of novel catalytic materials [1]. The promise of single atom catalysts is to lower the requirements for platinum group metals by utilizing these metals more efficiently and to create novel catalytic pathways. For industrial applications, single atom catalysts need to be stable under reaction conditions and demonstrate durability during accelerated aging. Recent research shows pathways for scalable synthesis of single atom catalysts that might deliver catalysts meeting the thermal durability requirements of industry while yielding reactivity improvements over conventional supported metal nanoparticle catalysts [2]. Since mobile single atoms constitute the dominant mechanism for catalyst sintering via Ostwald ripening, improving the stability of single atoms could help improve the durability of all heterogeneous catalysts used in industry. In this presentation we will describe recent work on an approach which we termed atom trapping [3]. Our initial work focused on trapping volatile metal oxides such as PtO_2, to improve the durability of Pt catalysts, but we are now learning how this approach can be more broadly applicable. We will describe how fundamental understanding of the stabilization of single atoms and sub-nanometer particles and clusters can be helpful in a range of applications ranging from emission control to energy conversion.

1. A.K. Datye and H. Guo, Single atom catalysis poised to transition from an academic curiosity to an industrially relevant technology. Nature Communications, 2021. 12(1): p. 1-3.
2. Regalbuto, J.R. and Datye, A.K., All the lonely atoms, where do they all belong? Nat Nanotechnol, 2022. 17(2): p. 110-111.
3. Jones, J., Xiong, H.F., Delariva, A.T., Peterson, E.J., Pham, H., Challa, S.R., Qi, G.S., Oh, S., Wiebenga, M.H., Hernandez, X.I.P., Wang, Y., and Datye, A.K., Thermally stable single-atom platinum-on-ceria catalysts via atom trapping. Science, 2016. 353(6295): p. 150-154.

51c Computational Materials Design with Machine Learning and Atomistic Simulations

Rafael Gomez-Bombarelli

Designing new materials is vital for addressing pressing societal challenges in health, energy, and sustainability. The computational techniques of atomistic simulation and machine learning (ML) offer an avenue to rapidly invent new materials and navigate this enormous space. By populating the continuum between physics-based simulations and machine learning, it is possible to enable rapid, computation-first design of materials that accelerate the materials discovery cycle.

Atomistic simulations, using techniques from quantum mechanics or statistical mechanics can predict the properties of hypothetical materials, and by engineering high-throughput simulation pipelines, millions of candidates can be evaluated to identify compositions or structures that optimize a given property. Simulations are, nevertheless, relatively costly, and may lack accuracy compared to experiment. This is where the synergy with ML enables a new paradigm: surrogate models bypass simulations by interpolating among pre-existing calculations at a fraction of the cost, while embedding physics-based priors in ML ensures robustness and transferability.

We will present our current progress in enabling end-to-end materials design for multiple materials classes and energy applications, from heterogeneous nanoporous catalysts, to crystalline and polymer solid electrolytes for batteries, to transition metal oxide for electrocatalysis.

30a Electrochemical Windows of Sulfone-Based Electrolytes for Lithium Metal Batteries: A Density Functional Theory and Cluster-Continuum Model Investigation

Minhee Park, JunBeom Cho, Bumjoon Seo, Won Bo Lee

Lithium metal batteries using lithium metal as an anode are attracting considerable interest as potential energy storage devices due to their high theoretical energy density. However, it still remains a technical bottleneck to find stable electrolyte materials that minimize the limit of selecting cathode or anode materials for high-voltage applications. Researchers report that sulfone-based electrolytes exhibit excellent oxidation stability, operating at electrochemical potentials ⟩ 5.0 V Li^+/Li. Sulfone-based electrolytes still suffer from distinct problems that must be addressed before commercial applications. One primary drawback is the occurrence of excessive side reactions between sulfones and Li metal, which limit the battery life cycle. Therefore, understanding the relationship between interactions in sulfone-based electrolytes and their redox properties is of critical significance in the battery industry.

Herein, the electrochemical stability windows of sulfone-based electrolyte systems were computed using both Molecular Dynamics (MD) simulation and Density Functional Theory (DFT) calculation. MD simulation was first conducted to identify the composition of solvation structures in the electrolyte system. Then, the full-scale electrochemical windows were generated by DFT calculations. According to the cluster-continuum approach, the first solvation shell coordinated with Li cation was explicitly defined while other solvent molecules were considered using an implicit solvation model. We compared the results of two electrolyte systems with different solvents, but identical salt anion of lithium-bis(fluoromethanesulfonyl)imide (LiFSI): ethyl methyl sulfone (EMS) and dimethyl sulfone (DMS) solvent based. The electrochemical stability windows were calculated for six main coordination structures to investigate the change in values depending on the structure, levels of theory, coordination number, solvation model, and solvent type. It was confirmed that the window width widened as the empirical terms were included in the exchange-correlation functional increased. By evaluating the redox reaction through Mulliken charge analysis, it was found that fluorine dissociation generated from salt anions increases the reduction potential by stabilizing the reduction state. Defining the solvent environment in an explicit manner, a key perspective is outlined for the electrolyte design of high-performance lithium metal batteries.

30b Uniformly Embedded Cobalt Phosphide Nanoparticles in Carbon Nanofibers As Sulfur Host Via Co-PBA in-Situ Growth for High-Performance Li-S Batteries

Zhi Liu, Cameron Romero, Kenneth Gordon, Le'Naedrea Moore, Ling Fei

Due to their high theoretical capacity and energy density, Li-S batteries have emerged as promising candidates for the next generation of commercial energy devices, attracting significant attention. However, existing challenges such as the shuttle effect of polysulfides and the inert conductivity of sulfur have hindered the development of these batteries. To overcome these challenges, researchers have recently explored the use of hierarchical carbon hybrid materials incorporating metallic compounds to improve the electrochemical performance of the cell. One such material, the use of prussian blue analogues (PBA) combined with carbon nanofibers as cathode materials has been reported rarely. In this study, a CoP/CNFs (cobalt phosphate/carbon nanofibers) composite was obtained by pyrolyzing electrospun fibers from PAN with in-situ grown Co-PBA, followed by phosphorization. The resulting CoP nanoparticles can efficiently promote the catalytic conversion of lithium polysulfide (LiPSs) due to a reduced energy barrier, effectively inhibiting shuttle-effect. The interconnected CNFs structure provides a high conductive framework and large exposed area, which accelerates electron transport and Li ions movement due to a reduction in accessing distance, and improvement in the absorption ability of LiPSs. The synergistic effect of the CoP nanoparticles and CNFs structure contributes to the high performance of Li-S cells, including cycling and rate capability. This promising strategy expands the routes for preparing cathode materials with high performance in Li-S batteries, overcoming existing challenges and improving the electrochemical performance of the cell.

30d High Resolution Measurement of Potential-Dependent Electrochemical Activities on HOPG Using Scanning Electrochemical Cell Microscopy (SECCM)

Myunghoon Choi, Sang-Joon Cho, Lane Baker, Stefan Kaemmer

HOPG is widely used as a substrate for scanning probe microscopy (SPM) research due to its flatness, uniformity, and ordered layered structure, making it ideal for various SPM techniques such as Scanning Tunneling and Atomic Force Microscopy. It is also commonly used as a model electrode substrate for developing and optimizing electrocatalytic materials and reactions, playing a vital role in energy conversion processes. HOPG's highly ordered structure and high electrical conductivity make it an excellent substrate for investigating electrocatalysis at the atomic and molecular level, providing insights into the fundamental mechanisms of electrocatalysis and designing more efficient catalysts for various applications. However, the HOPG surface exhibits heterogeneous electro- and electrocatalytic characteristics on its basal plane and edge due to differences in their electronic and structural properties.

Traditionally, scanning electrochemical microscopy (SECM) is employed to study the electrochemical heterogeneity at a surface, allowing for the visualization of electrochemical processes. SECM measures the electrochemical activity of local areas using a microelectrode probe, providing information on the electrocatalytic properties. However, the Unwin group has developed scanning electrochemical cell microscopy (SECCM), which offers higher spatial resolution measurements than SECM. SECCM achieves this by using a small meniscus formed at the pipette tip as a probe and electrochemical cell, enabling precise delivery of reactants to a specific location on the surface. The map of electrochemical activity measured by SECCM is correlated with AFM topography to identify the step edge from the basal plane, which exhibits higher electrochemical activity on the HOPG.

In this study, the heterogeneity in the electrochemical activities of HOPG was visualized using a customized SECCM setup that allowed for the simultaneous acquisition of topography and electrochemical activity maps. The HOPG grade 2 substrate used in this study possesses an intermediate quality suitable for both extensive use and high-resolution imaging purposes. A series of electrochemical maps at different potentials applied to the HOPG substrate were acquired by operating SECCM in AC-mode. Further details will be discussed in the poster.

30e Wetting Transparency of Free-Suspended Single-Layer Graphene on Liquid Substrate.

Fan Yang, Annette Thompson, Lei Li

Wetting transparency of single-layer graphene provides immense potential for the development of multi-functional devices in which the wettability of an atomic-thick film can be manipulated by selecting the appropriate supporting substrate. It becomes even more interesting when the substrate is liquid since flowing different liquids beneath the graphene will provide the real-time control of the surface wettability. However, the wetting transparency on liquid substrates has been rarely studied, possibly due to the challenges in performing experiments on liquid substrates. To this end, we have developed a contact angle measurement procedure to determine the surface and interfacial tension of free-suspended single-layer graphene on liquid substrates using Neumann's Triangle Model. By using different testing liquids on top of free-suspended single-layer-graphene or changing liquid substrate underneath single-layer-graphene, our results reveal that single-layer-graphene is nearly complete wetting transparent. This experimental approach can also be applied to other types of 2D materials for similar studies. Our findings provide insight into the underlying mechanism of wetting transparency on single-layer graphene and demonstrate the possibility of controlling the surface wettability of single-layer graphene by engineering the supporting substrate in real time. The results here have important implications in many applications, including water harvesting, microfluidic devices, and liquid-filled single-walled carbon nanotubes.

30f Feedback Control System of Iron Catalyst in Reactors Using Deep-Injection Floating Catalyst Chemical Vapor Deposition (DI-FCCVD)

Mingrui Gong, AnaVi Benavides Figueroa, Arthur Sloan, Matteo Pasquali

The current most promising process for scale-up of high-quality few-walled carbon nanotubes (FWCNTs) synthesis is the deep-injection floating catalyst chemical vapor deposition (DI-FCCVD) method. In our DI-FCCVD design, carbon feedstocks methane, hydrogen, and a catalyst carrying inert gas, are injected directly into the hot zone of a vertically oriented reactor. Vapor phase ferrocene and thiophene are used as catalyst precursors which react in situ to nucleate an aerosol of catalytic iron nanoparticles which serve as active sites for FWCNT growth.

Control over the catalyst particles and their formation is essential to the performance of the process, however the delivery of the catalyst precursors is poorly characterized and controlled in laboratory settings. Experimental catalyst delivery rates are often approximated using ideal bubbler assumptions of a perfectly saturated vapor, which may not be correct. To decrease the experimental variability, we have designed a feedback control system to provide an accurate and consistent ferrocene vapor delivery during synthesis using real-time acoustic measurement of ferrocene concentration. Using the process data provided by this control system we refined the design of our ferrocene sublimation system to counter sources of process variability that arise due to the characteristics of the powdered ferrocene used as a catalyst source. These sources of variability include sintering, the formation of channels, and powder entrainment in the gas flow. This design minimizes these factors while maximizing the contact of the carrier gas with the ferrocene, which remaining simple to build.

Implementing both the control system and the improved sublimator into our FCCVD reactor system has decreased run-to-run variability and allowed to begin quantitively studying how the catalyst is utilized in the system. By closing the catalyst mass balance by comparing the inlet iron flow rate to the iron recovered in the product we are able to characterize the effect of process changes on the loss of catalyst material and the efficiency of its utilization.

In this context, we alter the S/Fe ratio, injection depth, and velocities inside the injection tube to study their impacts on catalyst utilization and efficiency percentages. such percentages relationship with the methane conversion and production rate will provide information for more complex analysis of CNT formation conditions.

The improved process control provided by this system combined with the results of the quantitative catalyst studies has allowed us to exceptional CNT synthesis results. We can repeatably synthesize long (>10 μm), highly crystalline (Raman G/D > 50) FWCNTs with small diameters (1.42±0.06 nm), at high rates (~1 g/hr), with minimal residual catalysts (< 5 wt%) and non-CNT carbon content (>90% CNTs), while also achieving extremely high single pass methane conversion (>25%).

30g Facile Synthesis of Halogen Terminated Mxene from Different MAX Phases for Electromagnetic Interface Shielding

Shane Reed, Pitchaimari Gnanasekar, Sanjay Singh, Natalie Arnett, Dinadayalane Tandabany, Theo Siegrist, Subramanian Ramakrishnan

Two-dimensional (2D) layered metal carbides and nitrides are known as MXenes, with the versatile properties (high electrical and thermal conductivity) of 2D MXenes driving interest in various applications (batteries, fuel cells). Generally, MXenes are demonstrated by a general chemical formula of $M_{n+1}X_nT_x$ (n = 1–4), where M represents an early transition metal such as Ti, V, Mo etc.., where X is carbon/nitrogen, and T_x represents the surface terminations such as Br. Conventionally, MXenes are produced by the selective etching of the A interlayer from their parent three-dimensional (3D) MAX phase ($M_{n+1}AX_n$). Thus, the ability to modify the surface chemistry of MXenes is an important topic. However, an experimental investigation of new surface functional group termination on MXenes remains a challenge.

Over the past decade, an exhaustive amount of research has been conducted on the synthesis of MXenes via aqueous HF-based methods that leave an array of hydrophilic termination groups on the surface of the MXene, namely -OH, -O and -F. The serious toxicity issues raised by these compounds hinders the scale up production of MXenes. Moreover, different surface terminations, other than fluorine and hydroxide-based systems, are expected to broaden the applicable chemistries for the functionalization of MXenes. Recently, tuning surface modifications on MXenes were explored by solid-state methods such as molten salt etching. However, these methods suffer from several drawbacks; for example, the experimental conditions of the alkali- based etch are harsh making it hard for scale-up production, the molten salt method require relatively high temperatures (~550 °C), and finally the yields for the electrochemical synthesis approach are fairly low. Besides, regardless of the different etching routes available in the literature, MXene sheets suffer from uncontrollable surface terminations which inhibits the extent of possible functionalization. This has limited not only the production of high-quality MXenes but also the prospect of repeatable, theoretical studies in the literature to predict MXenes properties prior to use in many device applications.

Therefore, the potential of developing non-fluorine based etching procedures is of great interest as it will further development and mass production of MXenes for commercial applications. In this work, a novel non-aqueous, halogenated (bromine) etching from different MAX phases has been developed at room temperature to investigate the effect of surface chemistry on the properties of the MXene. Figure 1 illustrates that as etching proceeds, the color changes from dark red to light yellow over 24 hours at room temperature, indicating the consumption of Br_2, and it is evident that the halogen etching methods produce colloidally stable, multi layered MXene (Ti_3C_2Br) after processing. The XPS spectroscopy confirms the removal of the Al interlayer while preserving the titanium carbide backbone of the initial MAX phase (Fig. 2). Scanning Electron Microscopy is used to confirm the presence of exfoliated MXene flakes (Fig. 3). The reaction (bromine etching) on the MAX phase surface is monitored via wide-angle X-ray scattering by drop-casting aliquots from the reaction slurry onto silicon wafers (Fig. 4). A peak shift is clearly seen in figure 4 from 10 degrees to 7.5 degrees, indicating the presence of brominated MXenes (Ti_3C_2Br). The disappearance of the peak at 40 degrees indicates the removal of aluminum. The significance of this work is that it not only provides a complimentary, non-toxic, fluoride-free route to prepare MXenes, but also increases access to new chemistries for surface derivatization, which expands compositional-tunability of optical, electrical, and chemical properties. Applications of the halogen etch to different MAX phases will also be presented.

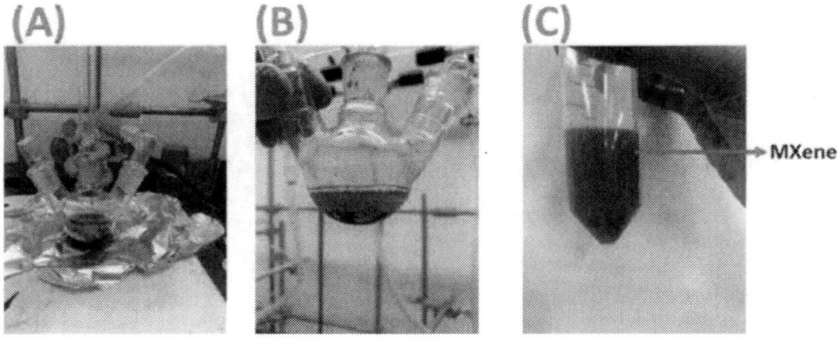

Figure 1. Schematic illustration of Br_2 Etch. A) Addition of Br_2 to MAX Phase (Ti_3AlC_2). B) After completion of the reaction (a pale-yellow supernatant indicating removal of Al interlayer). C) The colloidally stable MXene flakes are isolated from the un-etched MAX phase

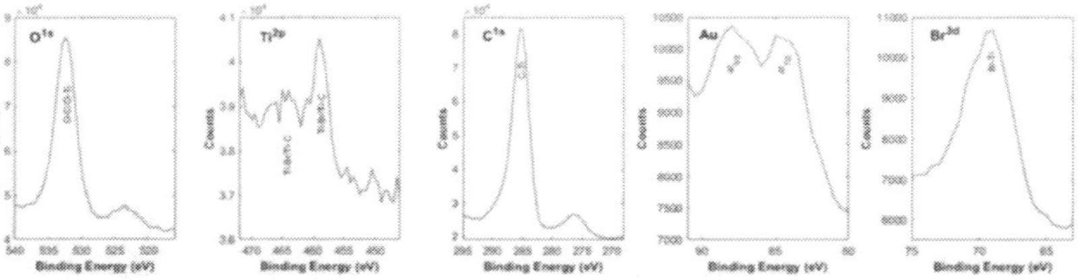

Figure 2. XPS of Bromine etched MXene (Ti_3C_2Br) detailing bonding characteristics

Figure 3. SEM of Bromine etched MXene (Ti_3C_2Br)

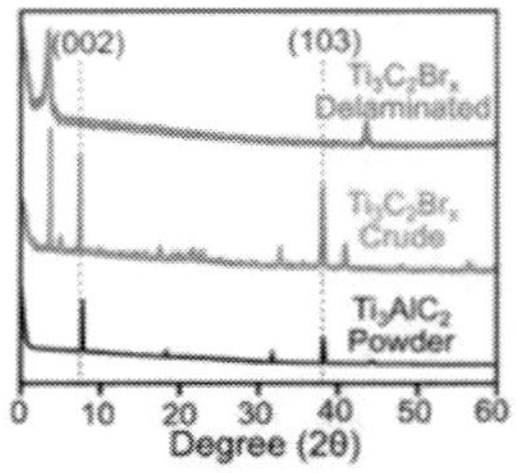

Figure 4. XRD of Bromine etched MXene (Ti_3C_2Br)

30h Rheo-Electric Investigation of Carbon Black Suspensions Undergoing Shear-Induced Microstructural Rearrangement

Paolo Ramos, Julie Hipp, Jeffrey Richards

Carbon black is ubiquitous as a conductive additive in electrochemically active slurries for emergent energy storage and capacitive desalination systems. When suspended, these high-structured, electrically conductive particles agglomerate, creating a charge-carrying microstructure that facilitates electron transport. In our initial work, we surface oxidized carbon black particles and enhanced their colloidal stability. We showed that reduced agglomeration led to a decrease in the electronic conductivities of the particles in deionized water. This demonstrated a non-trivial connection between the carbon black microstructure and its electrical response. This microstructure additionally evolves under flow, leading to a varied electrical response depending on the suspending medium and the flow conditions. To further investigate this varied response, we dispersed neat carbon black particles in Newtonian fluids with varying dielectric constants and viscosities. The formulated carbon black suspensions also spanned a range of volume fractions above the electrical and mechanical percolation thresholds for each solvent. We sheared the suspensions from 2500 to 10 s^{-1} and characterized the microstructural rearrangement of the agglomerates by calculating the Mason number. At each flow regime, we performed in situ impedance frequency sweeps on the suspensions and obtained their conductivities, relative permittivity, and dielectric strengths by fitting the dielectric spectra to several models. By parametrically pairing the measured mechanical and dielectric properties, we reveal the systematic dependence of the mechanism of charge transport on not only the microstructure of the carbon black agglomerates but also the solvent properties. We believe that this qualitative insight will provide crucial design rules towards the development of slurries with optimal performance.

30i Lyotropic Liquid Crystals of Cellulose Nanocrystal/Boron Nitride Nanotube Mixtures: Phase Behavior and Aligned Assemblies

Eren Katnas, Geyou Ao

Abstract:

Liquid crystals of one-dimensional (1D) boron nitride nanotubes (BNNTs) can be utilized to manufacture strong, multifunctional solid assemblies such as films and fibers with improved alignment and properties. Cellulose nanocrystals (CNCs) are known to form lyotropic liquid crystals and have been utilized as a structural template for nanomaterial assembly. Here, aqueous dispersions of BNNTs were stabilized by carboxymethyl cellulose nanofibers (CM-CNF) *via* noncovalent complexation. The individually dispersed BNNTs were then incorporated into aqueous dispersions of CNCs to obtain liquid crystals and the phase behavior of BNNT/CNC mixtures with increasing concentrations of BNNTs were studied using polarized light optical microscopy and rheology. Additionally, the wet spinning method was utilized to obtain solid fibers from BNNT/CNC mixtures with improved nanorod alignment and the resulting mechanical properties were measured. Fiber spinning from liquid crystals of BNNT/CNC mixtures can potentially better translate the alignment and properties of nanorods to macroscopic assemblies, which will lead to developing BNNT-based applications in many areas ranging from thermal management to multifunctional textiles.

30j Scalable Manufacturing of Inorganic Nanomaterials Using Jet Mixing Reactors

Priya Jana, Faiz Khan, Pinaki Ranadive, Jessica Winter, Nicholas Brunelli

Typical laboratory batch reactors for inorganic nanoparticle synthesis can be difficult to scale since rapid particle nucleation and growth require efficient mixing to produce monodisperse particle size distribution (PDI). As the size of the reactor increases, it becomes increasingly difficult to achieve uniform temperature and concentration, leading to variations in the size, shape, and composition of the nanoparticles. These factors limit the use of inorganic nanoparticles in commercial applications ranging from energy to healthcare. Continuous microreactor-based synthesis is an attractive alternative, providing a uniform particle size distribution with mixing times of a few milliseconds. This work explores a novel jet-mixing reactor (JMR) design consisting of an axial flow with two jets impinging on the mainline, resulting in a single stream that exits the reactor. Previous work in our laboratory successfully synthesized metal nanoparticles and inorganic nanocatalysts like Pd@TiO_2 in JMR at room temperature using Palladium nanoparticles pre-synthesized in batch. The current ongoing research builds on this by integrating batch methods into a single continuous process using JMRs in series to produce Pd@TiO_2 which requires balancing flow rates and heating/cooling rates. Because Palladium nanoparticles are sensitive to reaction temperature, we are investigating the construction and characterization of the jet mixing reactor at high temperatures through nanoparticle synthesis and scaling analysis. Palladium-catalyzed reactions for biomass upgrading will then be further studied to quantify the yield of these heterogeneous nanocatalysts. Mixing time and size distribution studies will then be further explored to address scalability issues.

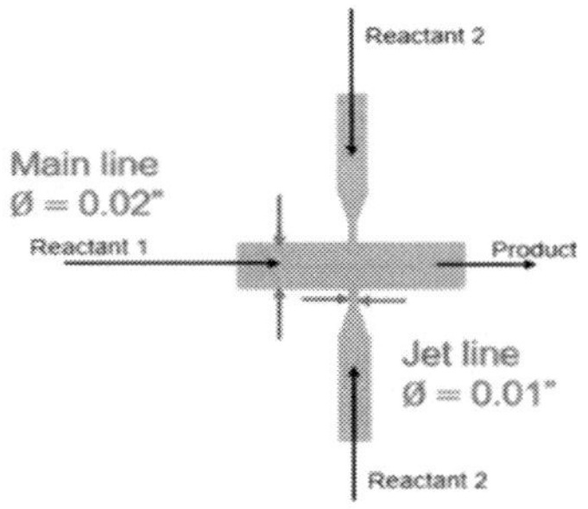

Figure. Reactor design with two jet lines impinging on the main line producing a single outlet for the product.

30k Threshold and Excess Electrolyte-to-Sulfur (E/S) Ratios in Lean-Electrolyte Li-S Batteries

Xiaosi Gao, Siddharth Tiwari, Aditya Pandey, Yiqi Shao, Ziang Gao, Yong Lak Joo

Lithium-sulfur (Li-S) batteries have been recognized as one of the most promising next-generation alternatives to Li-ion batteries (LIB). Specifically, lean-electrolyte Li-S batteries are receiving more attention for lower dead weights, and thus higher overall energy densities. It was calculated that the electrolyte-to-sulfur (E/S) ratio should be < 5uL/mg for Li-S batteries to be comparable with the state-of-art commercial LIB. Many efforts have been put into the development of lean-electrolyte Li-S batteries from all aspects (cathode, anode, separator, current collector...) and have achieved great improvements. However, most articles focused on one fixed condition of E/S ratio and sulfur loading, and few have performed a predictive analysis of Li-S battery performance in a wide range of both E/S ratios and sulfur loadings.

In this study, we analyzed the initial discharge capacity and cycling stability of Li-S batteries at E/S ratios of 2~30 uL/mg, from which we proposed a quantitative correlation. Interestingly, at lean electrolyte conditions < 10uL/mg, we found discharge capacity not proportional to E/S ratio, as expected from the polysulfide solubility. At each mass loading, there exists a threshold E/S ratio for the battery to function properly, possibly due to wetting of separator and electrode. We also determined the "excessive" E/S ratio signifying the "flooding" conditions, above which the capacity is no longer limited by amount of electrolyte. Both threshold and excess E/S ratios were found to be dependent on sulfur mass loadings, which were determined experimentally at 1~5 mg/cm^2. Our study was also extended to pouch cells to further approach the industrial standard, during which we found the same trend. We demonstrate our results can provide some insights into the design of lean-electrolyte high-sulfur-loading Li-S batteries which can push Li-S batteries further into practical commercialization.

301 Glycopolymer-Wrapped Carbon Nanotubes Detection of Viral Proteins

Brandon Heppe, Geyou Ao, Xue-Long Sun

There is a critical need for establishing versatile and effective viral detection platforms since the global pandemic of SARS-CoV-2 and the initial limitations in testing capabilities. Carbohydrate-mediated interactions play a key role in many cellular events, including viral and bacterial infection, that is facilitated by selective interactions of carbohydrates with carbohydrate-binding proteins (CBPs). Glyconanomaterials with specific carbohydrate functionalities are known to mimic the multivalent carbohydrate-protein interactions occurring in biology. Our previous work has investigated the specific interactions of glycopolymer-wrapped single-wall carbon nanotubes (Glyco-SWCNTs) with CBPs using lectins. Specifically, Glyco-SWCNTs were created *via* noncovalent complexation with lactose-containing glycopolymers. Here, we build upon our previous findings to optically detect the specific binding of SARS-CoV-2 spike protein to carbohydrates by utilizing the intrinsic near-infrared fluorescence of Glyco-SWCNTs. Ultimately, this work will lead to the development of microbial nanosensors and further enable the mechanistic understanding of carbohydrate-protein interactions occurring in pathogen infection.

30n Polymer-Cerium Oxide Nanocomposites Synthesized By Hydrothermal Method for Bone Tissue Regeneration Applications

Shreya Pawar

One common detrimental effect on biological system caused by orthopedic pathogenesis is bone tissue degeneration. Relevant orthopedic pathogeneses include bone cancer and age-related illnesses, such as osteoporosis and osteopenia. One major cause of bone tissue degeneration is the high generation of reactive oxygen species (ROS) associated with many related pathologies. The accumulation of ROS stimulates the activities of osteoclasts, while inhibiting the activities of osteoblasts. Thus, overall there is a significant loss in bone tissue. Additionally, the accumulation of ROS also hinders the activities of naturally-occurring enzymes such as superoxide dismutase (SOD) and catalase (CAT) that normally function to catalyze the removal of ROS from biological tissue. Bisphosphonates are currently used in medical practice to slow bone tissue degeneration and facilitate healing. However, bisphosphonate therapeutic efficacy is compromised due to short circulation times, limited bioavailability to target tissues, and uncontrolled release by current methods. To overcome these limitations, we have synthesized polymer-cerium oxide nanocomposites (pCNP) using sacrificial template materials. Sacrificial template materials allowed us to control the size of the porous cerium oxide nanoparticles. Recent studies show pCNPs formulations have high ROS scavenging abilities to mitigate bone degeneration. This property can be attributed to the redox potential of the Ce^{3+} state and surface SOD-mimetic and CAT-mimetic properties of polymer cerium oxide nanocomposites. Synthesizing pCNPs increases nanocomposite surface area and drug encapsulation relative to conventional cerium oxide nanoparticles. To optimize particle specific surface area and surface defect morphology for optimally efficient drug loading, we have experimented with the following different templates: silica, agarose, glucose, polyvinyl alcohol (PVA), polyvinylpyrrolidone (PVP), and cetrimonium bromide (CTAB). Each templated synthesis had its own variability in size and offered unique characteristics to the polymer cerium oxide nanocomposites. pCNPs were synthesized *via* a hydrothermal method, to promote cation hydrolysis and oxide crystallization. These chemical processes yielded higher surface area through maximization of surface defects. Successful synthesis of the polymer-cerium oxide nanocomposites was confirmed through characterization studies such as thermogravimetric analysis, FTIR, UV-Vis, DLS size analysis, and TEM imaging. Surface area (BET pore size analysis) measurements were performed to investigate pCNP specific surface areas for formulations produced from different templates to determine optimal synthesis procedures. Superoxide dismutase and catalase enzyme-mimetic activity assays confirmed pCNP formulations abilities to interact with ROS and enhance their clearance: facilitating bone tissue regeneration. MTT cell viability assays showed pCNP's biocompatibility with human mesenchymal stem cells. These cell viability assays showed that 10 ug/mL concentration was the ideal biocompatible concentration. Furthermore glucose-cerium oxide nanocomposites and agarose-cerium oxide nanocomposites yielded the best cell viability over a 7 day period. CTAB-cerium oxide nanocomposite did not show to have the optimal biocompatibility with human mesenchymal stem cells. However, overall for all synthesized polymer-cerium oxide nanocomposites, cell viability over 7 days exceeded 80%, indicating that our formulations had no cytotoxicity on human mesenchymal stem cells. Studies were also performed to determine if pCNPs can induce human mesenchymal stem cell differentiation into osteoblasts bone stem cell. This was assessed through the Alkaline Phosphatase (ALP) and Alizarin Red (ARS) assays. ALP assays, quantification, and staining showed that the synthesized polymer-cerium oxide nanocomposites have promising potential to induce human mesenchymal stem cell differentiation into osteoblasts. The ALP assays showed that glucose-cerium oxide, agarose-cerium oxide, and PVP-cerium oxide nanocomposites had the best ability to facilitate mesenchymal stem cell differentiation. Currently, further tests through the ARS assay are being conducted to determine the polymer-cerium oxide nanocomposite formulations ability to stimulate calcium deposition and bone formation. Overall, comprehensive studies show pCNP's capability to facilitate bone tissue regeneration in orthopedic diseases. Thus, polymer cerium oxide nanocomposites show great potential to provide a synergistic effect with conventional clinical bisphosphonate to amplify therapeutic efficacy.

30q Lipid Nanoparticle Structure and Immune Response Regulate mRNA Organ Tropism and Pup Growth during Pregnancy in Mice.

Namit Chaudhary, Alexandra Newby, Mariah Arral, Saigopalakrishna Yerneni, Samuel LoPresti, Rose Doerfler, Daria Petersen, Bethany Fox, Tiffany Coon, Angela Malaney, Yoel Sadovsky, Kathryn Whitehead

Treating pregnancy-related disorders is challenging because many relevant medications cause maternal and fetal toxicity. This potential danger has hindered drug development for pregnancy-related disorders for several decades. Lipid nanoparticle (LNP)-based RNA therapies with high delivery efficacy and favorable safety profile can quell toxicity concerns. To this extent, we report LNP structures that effectively deliver mRNA to maternal organs and placenta with no fetal delivery. In the placenta, LNPs transfect trophoblasts, endothelial cells, and immune cells. Additionally, we show that immune response during pregnancy alter LNP organ tropism and hinder pup growth. Together, our results identify potent LNP structures and design criteria for safe mRNA delivery during pregnancy.

30r The Particle Drifting Effect – Impact of Membrane, Sink Condition, Particle Size, Drug Solubility, and Colloid Type

Na Li

Colloidal drug particles, such as amorphous drug aggregates, nanocrystals, and bile micelles, are known to promote oral absorption of poorly soluble drugs. As only solubilized drug can permeate through membranes freely, it remains less well understood how and to what extent these colloidal particles contribute to drug absorption. Thus, the aim of this study is to gain insights on the mechanisms and extent of enhanced absorption by colloidal particles. Various experimental setups, including artificial membranes, biphasic diffusion, cell culture models, and animal studies were used to investigate this phenomenon, and colloidal particles of different size, type, and compositions were tested. Our results suggested minimal contribution from endocytosis, whereas passive permeation is the main mechanism of enhanced absorption by drug nanoparticles. Colloidal drug particles dissolve or dissociate, and replenish the consumed free drug concentration within the diffusional boundary layer, thus leading to increased local concentration in the aqueous solution adjacent to the membrane surface. The extent of such enhancement was found to be dependent on both the particle dissolution/dissociation and membrane permeation processes, where colloidal and drug properties as well as the relative contribution of diffusional boundary layer appeared to be key factors. Mass transport analyses were performed to provide more quantitative understanding of this phenomenon. This work provides insight on the permeability advantages by drug particles and colloidal species, and may contribute to improved bioavailability prediction of colloid containing formulations.

118a Biodegradable Polymer Capsule for Long-Term Drug Release

Zhiwei Zheng, Thomas Waterkotte, Tilahun Debele, Gregory Dion, Yoonjee Park

Voice disorders, which are common communication disorders in the US, can result from various conditions affecting the larynx, respiratory system, or vocal tract. Vocal fold scarring, often caused by injury and inflammation, is a common cause of dysphonia, and current therapeutic options are limited. Airway stenosis, a narrowing of the airway, poses a serious problem for both adults and children and can result from various causes. Surgical procedures are often required to address severe cases of airway stenosis, but they can lead to inflammation and other complications.

Corticosteroids, known for their anti-inflammatory properties, have been used in the treatment of laryngeal diseases. Local administration of corticosteroids directly into the larynx has shown promising results, especially in reducing scarring and inflammation. However, the effectiveness of current corticosteroid delivery methods is limited by factors such as short duration of action, discomfort, and costs associated with repeated procedures.

To address these limitations, we have developed sustained-release capsule implants for local delivery of dexamethasone (Dex). Four different biodegradable polymers (polycaprolactone, poly(lactic acid), 90:10 poly(lactic-co-glycolide), and 50:50 poly(lactic-co-glycolide)) were used as capsule materials. The release profiles of Dex from these capsules were compared, and the 50:50 poly(lactic-co-glycolide) showed the fastest release. The permeability and partition coefficient of each polymer were measured, and it was found that the release of Dex through the capsule membrane was primarily due to "solution-diffusion" rather than diffusion through pores. The hydrophobicity of the polymer sheet was found to correlate with the drug release kinetics, with more hydrophobic polymers exhibiting slower release rates for the hydrophilic drug Dex. In vitro studies demonstrated the effective suppression of pro-inflammatory cytokine IL-6 levels using the Dex-loaded implant, indicating its potential therapeutic benefits with minimal toxicity.

Overall, the development of the biodegradable polymer capsule for long-term release of Dex holds promise for improving the treatment of voice disorders, airway stenosis, and other related conditions.

118b Magnetic Nanocomposites for Environmental Remediation

Pranto Paul, Marissa Nicholson, J. Zach Hilt

Magnetic nanocomposites can provide unique solutions to environmental problems. The magnetic nanoparticles embedded in nanocomposite enable energy to be delivered remotely to the material using an alternating magnetic field (AMF). This localized energy can be used to drive processes to enhance material performance, including disruption of binding interactions, increased reactivity etc. Here, we will highlight a few diverse applications (e.g., sorb/desorb contaminant, activating radical, degrading contaminant). For example, decades of use of halogenated chemicals in product manufacturing and commercial applications have been connected to detrimental health impacts. In response to these, we have applied magnetic nanocomposite for the sorption of contaminants such as perfluoroalkyl substances (PFAS) and desorption through an AMF.

118c Upcycling of Waste Polyvinyl Chloride (PVC)-Based Materials to Electrospun Nanofibers for Water Treatment

Atta Ur Razzaq, Milad Esfahani

In recent decades, the increased use of polyvinyl chloride (PVC) in industries and households has led to a surge in PVC waste pollution, which mandates developing solutions for the removal of waste PVC from the environment. We report upcycling, the conversion of waste material to a high-value-added product, of PVC-based products to electrospun fibers (membranes). Two common PVC items, waste PVC pipe and waste PVC pool floats, were chosen and upcycled into electrospun membranes for water treatment. The fabrication process and fiber characteristics, such as morphology, surface charge, and mechanical strength of upcycled fibers, were studied and compared with the same fibers fabricated using research-grade (RG) PVC (commercial PVC powder). In addition, the effect of additives such as calcium carbonate in PVC waste products on the physicochemical properties of upcycled fibers was evaluated. The results showed that upcycling of waste PVC to electrospun fibers is feasible since the upcycled fibers showed similar or superior properties compared to their equivalent fibers from RG-PVC. Finally, the performance of upcycled fibers on the removal of dyes from the water was evaluated. The upcycled fibers from waste PVC pipe and pool float outperformed the RG-PVC fibers in removing methylene blue (MB) from water by showing more than 97 % removal efficiency and adsorption capacity of 195 mg/g compared to RG-PVC with 89 % removal and 179 mg/g adsorption capacity. This study clearly shows the potential of upcycling PVC waste materials from landfill into membranes for enhanced water treatment as a promising solution to the current plastic and water pollution.

118d Design and Properties of Magnetic Particle Imaging (MPI) Tracers Using Nitrodopa Anchored Coatings.

Neel Eswaran, Hayden Good, Ambar Velazquez Albino, Eric Imhoff, Carlos Rinaldi-Ramos

Magnetic particle imaging (MPI) is an emerging biomedical imaging technology that involves the use of iron oxide nanoparticles as tracers. Colloidal stability in biological environments is crucial to maintain resolution and sensitivity of tracers engineered for MPI applications such as blood pool imaging and cell tracking. Common methods of synthesis of high-quality iron oxide nanoparticles yield hydrophobic particles that are not suitable for such applications. A large body of work has explored phase transfer methods previously in the context of applications such as hyperthermia and magnetic resonance imaging contrast. However, MPI tracer design requirements differ from these applications in terms of size and magnetic properties of the inorganic core. As these properties affect the strength of magnetic attraction, it is important to re-evaluate methods of phase transfer and nature of particle coatings to achieve colloidal stability for tracers engineered for high sensitivity and resolution in MPI. Here, we report on the design and characterization of superparamagnetic iron oxide nanoparticle tracers optimized for MPI and phase-transferred to aqueous media using a 6-nitro-L-3,4 dihydroxyphenylalanine (nitroDOPA) anchor group conjugated to a polyethylene glycol chain (PEG-nitroDOPA). The nitroDOPA anchor group was selected due to its reported high affinity for the surface of iron oxide nanoparticles resulting in practically irreversible binding. At present, we have successfully synthesized the PEG-nitroDOPA polymer. We have also successfully stabilized iron oxide nanoparticles synthesized in our lab in aqueous media using the synthesized PEG-nitroDOPA, thus showing that despite the difference in diameter between the nanoparticles used in MPI (20-25 nm) and those reported in literature (typically 8-10 nm), the colloidal stability of the iron oxide nanoparticles is achieved. Dynamic light scattering (DLS) measurements of the iron oxide nanoparticles stabilized in water show a hydrodynamic diameter of less than 100 nm. Ongoing work seeks to reduce this further to ~50 nm range while preserving stability in biological environments and maintaining MPI sensitivity and resolution and evaluate blood circulation lifetime for the particles using non-invasive MPI assessment.

118e Automated Fe Quantification Assay to Accelerate Magnetic Nanoparticle Studies

Eric Imhoff, Ambar Velazquez Albino, Esteban Bermudez-Berrios, Carlos Rinaldi-Ramos

Iron concentration is an important parameter used in determining many characteristics of iron oxide nanoparticles. This includes characteristics that require a known mass of iron oxide to determine, such as nanoparticle concentration or saturation magnetization, and measurements that need to be normalized for consistency, such as signal intensity in magnetic particle imaging and in hyperthermia. Iron quantification assays are commonly performed using UV-vis spectroscopy. Typically, nanoparticle samples are mixed with various reagents in series to digest the iron to an aqueous salt before forming a colored complex in solution that allows absorbance measurement. These assays tend to be time-consuming, typically taking several hours and requiring significant manual effort as many repetitive pipetting steps are involved, especially when processing many samples at once. This study aims to evaluate the performance of a liquid handling robot as a replacement to manually performed assays with the goal of reducing time spent performing menial labor without sacrificing accuracy or precision of results.

A two-step protocol was designed to be used with the Opentrons OT2 liquid handler. Early procedures using nitric acid to digest the samples in the robot resulted in undesirable rapid corrosion of many parts which may reduce equipment longevity and introduce contamination. To prevent this and to address problems beyond the capability of the robot, such as particle sedimentation and phase separation during mixing, the first step of overnight particle digestion in acid was performed manually. Following the transfer of digested samples to a quartz 96 well plate, the robot performed all subsequent pipetting and mixing steps required using a multichannel pipette for increased throughput. 3D printing was also utilized to allow for rapid design and prototyping of custom equipment such as tube racks and reservoirs. This has reduced the person-hours required to complete the assay significantly without any appreciable loss in accuracy. One test with 16 identical samples of commercial particle Ferucarbotran resulted in a measured iron concentration close to the value measured in a manual assay with a standard error of 5%. The protocol has been validated with aqueous and organic solvents and particles of various coatings including oleic acid, polyethylene glycol, and poly (maleic anhydride-alt-1-octadecene). Future work will focus on improving the efficiency of the protocol by minimizing the human interaction required and expanding the liquid handler to other procedures.

Figure 1: Liquid handling robot with equipment for iron concentration assay.

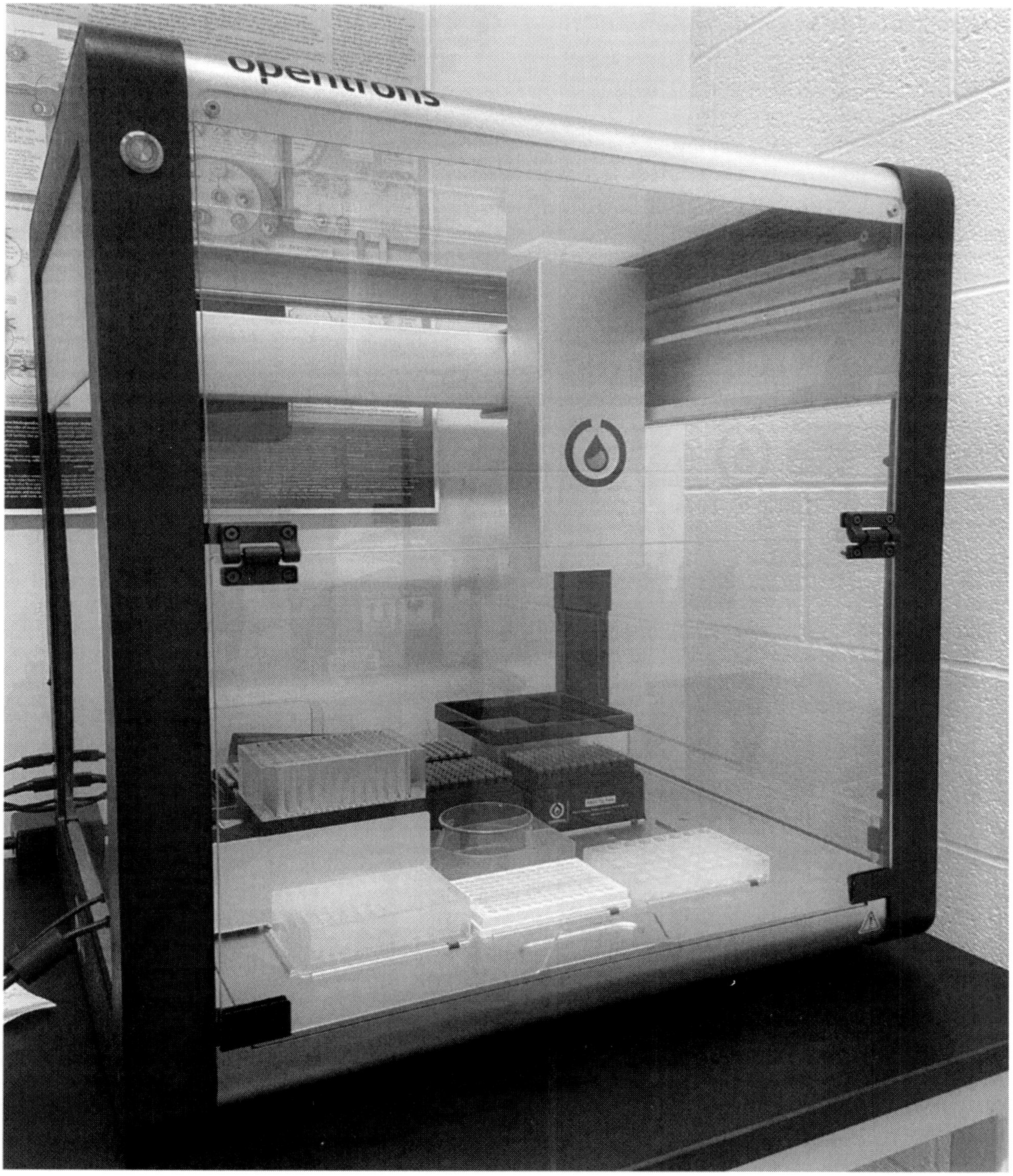
opentrons

118f Evaluating the Quality Control of Aqueous Carbon Nanotube Dispersions Using UV-Vis Spectroscopy and Zetasizer Analysis

Victoria Stege, Chelsea Monty-Bromer, Shelby Daniels, Michael Fulmer, Zachary Cheney

Currently, researchers are exploring many options to reliably produce and evaluate the quality of carbon nanotube dispersions for manufacturing applications. However, it can be a challenge to determine the quality of these large-scale dispersions quickly and accurately for use in a manufacturing setting. In this research, we will explore the evaluation of multi-walled carbon nanotube (MWCNT) dispersions using a UV-Vis spectrophotometer and a Zetasizer Nano for the creation of a nanofiber/MWCNT sensor. During this work, we determined the quality of MWCNT dispersions that would be suitable for use at a manufacturing site using UV-Vis and a Zetasizer Nano. Carbon nanotube dispersions were produced using bath sonication for 60 minutes or probe tip ultrasonication for 2 or 4 minutes at 24 kHz. These aqueous dispersions were created using 4:1, 6:1, and 10:1 surfactant to MWCNT ratios. The surfactants used for the dispersions were Triton X-100 and Triton X-114. Samples at a concentration of 1:50 (MWCNT dispersion to water) were taken from the dispersion and run through a Shimadzu UV-1800 Scanning Spectrophotometer and a Malvern Zetasizer Nano ZS to determine their quality. Based on preliminary results from the differing types of dispersions, the probe tip ultrasonication with 10:1 Triton X-100 to MWCNT ratio produced the most consistent dispersions.

118g Surfactant Mediated Phase Transfer of Iron Oxide Nanoparticles

Esteban Bermudez-Berrios, Ambar Velazquez Albino, Eric Imhoff, Carlos Rinaldi-Ramos

Magnetic nanoparticles are an attractive platform for imaging and drug delivery. Magnetic Particle Imaging (MPI) is a novel imaging modality that detects the signal of superparamagnetic iron oxide nanoparticles (SPIONs) as they respond to an applied magnetic field. The signal is quantitative, proportional to the iron mass, unambiguous, and unattenuated by tissue depth, which is an advantage compared to traditional imaging procedures. Designing SPIONs whose surface properties make them colloidally stable in tissue culture media and biocompatible with T cells will allow their use to track immune therapies *in vivo*. Adoptive cellular therapy (ACT), for example, is a type of immunotherapy that uses a patient's own cells to target and destroy cancer cells. Clinical studies have suggested that the effectiveness of ACT in solid tumors is dependent on the accumulation and persistence of T cells at the tumor site. Therefore, using MPI to non-invasively track T cells will enable visualization of their distribution within a sample and help understand the mechanisms of action and failure of ACT immunotherapies.

Many of the methods used to synthesize nanoparticles develop them in organic solvents. The thermal decomposition of iron oleate used in our lab yields SPIONs with consistent physical and magnetic properties, but in a complex mixture of organic solvents and surfactants that provide stability under those conditions. Through a series of washes, most of the organic compounds can be removed, leaving behind hydrophobic SPIONs coated with a surfactant monolayer. However, SPIONs for biomedical applications need to maintain colloidal stability in aqueous media, as biological systems are mostly composed of water. The phase transfer of SPIONs from organic to aqueous media can be achieved with the help of surfactants which coat the surface to facilitate the effective transfer to water. Given the wide variety of surfactants available and the broad range of processing conditions that can be used to achieve phase transfer, the composition and process design parameter space is very large and high-throughput approaches are desirable.

Our goal is to develop a platform for high-throughput evaluation of surfactant double layer phase transfer compositions and processing conditions. In this study, we report initial progress using oleic acid (OA) to mediate the phase transfer of SPIONs from organic solution into water. Since OA is already present on the surface of as synthesized nanoparticles, the expectation is that additional OA will form a surfactant double layer (SDL). The SDL arranges in such a way that the hydrophobic tails interdigitate, and the hydrophilic heads are adsorbed on the SPION surface on one layer and interact outwardly with water molecules on the other layer. Phase transfer is achieved by vigorously mixing the organic phase containing the SPIONs with a volume of deionized water. This is followed by tip ultrasonication, which causes an oil-in-water emulsion to form, so that more of the SPIONs are in contact with water rather than aggregated within the oil droplets.

Initial studies suggest successful transfer of SPIONs to water. While the average SPION size observed using transmission electron microscopy (TEM) is ~20 nm in the organic solvent, the hydrodynamic sizes of the particles are much larger after transfer into water, suggesting aggregation. A positive relationship between addition of surfactant and improved transfer yield percentage has been observed. Conversely, our data suggests that lower amounts of OA achieve phase transfer of single particles. Ongoing work seeks to optimize phase transfer yield and reduce hydrodynamic size, evaluate stability of the resulting particles in T cell culture media, and evaluate phase transfer using combinations of surfactant to tune particle zeta potential, stability, and interaction with T cells.

118h Corona Phase Molecular Recognition of the Interleukin-6 (IL-6) Family of Cytokines Using Nir Fluorescent Single Walled Carbon Nanotubes

Xiaojia Jin, Xun Gong, Volodymyr Koman, Daniel James Lundberg, Michael Strano

The Interleukin-6 (IL-6) family of cytokines regulate inflammation and play important roles in numerous biochemical pathways. Understanding the fundamental molecular recognition of these molecules with synthetic substrates is key to new biomedical technologies, including assays, sensors, therapeutics and purification methods. Herein, we use the Corona Phase Molecular Recognition (CoPhMoRe) approach to engineer new carbon nanotubes constructs and study their subsequent binding to the inflammatory cytokines: IL6, interleukin-11 (IL-11), ciliary neurotrophic factor (CNTF) and leukemia inhibitory factor (LIF). Library screening identified two polymer–based CoPhMoRe constructs consisting of single-walled carbon nanotubes complexed with p(AA_{68}-*rand*-BA_{16}-*rand*-CD_{16}) polymer (MK2) or p(SS_{80}-*rand*-BS_{20}) polymer (P14) corona phases. The resulting dissociation constants (K_D) were 8.38 ng/mL and 16.7 μg/mL respectively, compared to that of the natural IL-6 receptor at ~0.32ng/mL. Comparative binding experiments suggests that both constructs appear to recognize the axially aligned alpha-helical structures present in the Interleukin-6 family. The findings from this study elucidate how nanoparticle interfaces can be designed to lock onto specific protein features, particularly alpha-helical structures, to enable molecular recognition and new types of sensing technologies.

118i Rational Synthesis, Morphology, and Propertyrelationship of Non-Precious Bi-Metallic Hybridnanocomposites for Improving Oxygen Reductionreactions

Mahshid Mokhtarnejad, Mariana Milano-Benitez, Bamin Khomami

Tailored metal oxide/metal-organic frameworks (MOFs) can readily be converted to hybrid nanocomposites (HNCs) comprised of transition metals interfaced with carbon (C) matrices. To that end, this class of nanocomposites (NCs) that exhibit superior conductivity and high porosity are promising and economically viable candidates as oxygen reduction reaction (ORR) electrocatalysts. Hence, optimizing the performance of this class of material by tailoring their composition, structure, size, and morphology through an efficient synthesis process has increasingly become an active area of research. In this study, we report the use of Laser Ablation Synthesis in Solution in Tandem with Galvanic Replacement Reaction (LASiS-GRR) technique as a facile, environmentally friendly, and rapid route for the synthesis and control of complex hierarchical bimetallic Zeolitic Imidazolate Framework (ZIF) structures, and their pyrolytic post-processing to yield metal oxide/MOF-based functional HNCs as highly active ORR electrocatalysts. The LASiS-GRR technique utilizes high-energy laser plasma reactions as the rate-controlling step for the initiation of solution-phase galvanic reactions. The one-pot, two-step process introduced in this study allows tailoring of the composition, structure, size, and morphology of the Co-based MOFs encapsulated in Zn-based porous MOF (Co/ZIF- 67@Zn/ZIF-8) crystals. Pyrolytic post-processing of these crystals leads to the creation of HNCs with superior functionality for ORR electrocatalysts for anion exchange membrane fuel cells (AEMFCs). Specifically, herein, for the first time, we demonstrate the use of the LASiS-GRR technique as a one-pot rational synthesis method for synthesizing Co/ZIF-67@Zn/ZIF-8 bi-MOFs that, upon post-pyrolytic treatments are converted into unique tailored HNCs of ZnO-interfaced ZIF-67 in carbonaceous matrices (ZnO/ZIF@C). Not only the effect of two distinct salt precursors on the particle size and morphology is investigated for the first time during LASiS, but also the superior performance and stability of these ZnO/ZIF@C HNCs as non-precious metal ORR electrocatalysts compared to commonly used Pt- and platinum-group metals (PGM)-based electrocatalysts in AEMFC applications is demonstrated.

319a Using Liquid Crystals to Control Molecular Packing in Organic Solids over Macroscopic Dimensions

Kushal Bagchi, Paul F. Nealey

The growth of single crystals is the traditional route to control molecular structure over macroscopic distances (~ 1cm). However, single crystals cannot be grown for all molecules. Moreover, crystals usually exhibit poor mechanical properties and limited compositional flexibility. Organizing and solidifying liquid crystals provides an alternative to traditional crystallography. We discuss here the use of nanofabrication to direct the self-assembly of blue-phase and discotic liquid crystals. Blue-phases are cubic liquid crystalline phases observed in systems with a high degree of chirality. We demonstrate how aligned blue-phases, prepared using nanolithography, can be used to prepare single crystalline, *polymeric photonic crystals*. The polymeric single crystals we prepare could be of broad interest for sensing and photonic applications. We also discuss the directed self-assembly of discotic liquid crystals, which are mesophases observed in systems with an aromatic "disk-like" core and alkyl side chains. We demonstrate the alignment of discotic liquid crystals in lithographically defined microchannels, and the growth of biaxial crystals from the aligned mesophases. As discotic liquid crystalline phases are observed among important organic semiconductor classes such as Porphyrins, Phthalocyanines and Hexabenzocoronenes, we expect our results to be of broad interest for organic electronic applications. Finally, future directions are discussed for harnessing liquid crystallinity to prepare novel organic materials, such as anisotropic polymers and molecular glasses. The proposed work aims to create solids that combine the most desirable properties of crystalline and amorphous solids and could have broad implications for applications such as organic electronics, photonics, and sensing.

319b Exploring the Potential of Smart Supramolecular Materials for Energy and Beyond

Shuhao Liu, Yu-Ting Lin, Bhargavi Bhat, Mustafa Akbulut

Supramolecular complexes, which are held together by non-covalent interactions, offer unique physicochemical properties that are distinct from their constituent molecules. Recently, the development of smart supramolecular complexes has become attractive due to their adjustable and controllable properties in response to internal or external stimuli. These intelligent materials have advantages in various fields, including energy, biology, medicine, horticulture, and others.

In the energy field, the development of smart supramolecular systems has the potential to enhance the production of fuel from traditional and unconventional reservoirs at minimum cost and high productivity. Our research focused on the development of pH-stimuli and temperature-triggered supramolecular complexes based on surfactants and hydrotopes, which can be controlled in a facile manner and respond swiftly, while being economically and environmentally friendly. For these systems, a maximum ~ 16,000-fold adjustability of viscosity could be achieved by adjusting the pH and temperature. Non-Maxwellian and Maxwellian fluid behavior transition could be controlled through changing the environment conditions. The molecular and micellar structure change contribute to these outstanding rheological properties. Stud

The potential applications of these smart supramolecular complexes were tested as candidate additives for enhanced oil recovery and hydraulic fracturing to improve oil and gas production. These intelligent materials can also be beneficial for other fields, such as personal care, lubricants, and biomedical applications, due to their intriguing properties. Overall, smart supramolecular complexes have promising potential for various applications in different fields, including the energy industry.

319c Holistic Understanding of Nanocrystal Stability

Dr. Aniket Sandip Mule

At a very small-scale regime (< 10 nm), the properties of materials strongly depend on the size, shape, and surface of the crystals. Consequently, nanocrystals exhibit functionalities that are markedly different from their bulk counterparts. The broad tunability of the properties at the nanoscale combined with an enhanced surface-to-volume ratio makes nanocrystals highly relevant to various applications ranging from optoelectronics to catalysis to medicinal chemistry.[1] However, nanocrystals often exhibit poor stability compared to bulk materials, thus, limiting their capabilities in all the applications mentioned above. Overcoming such a challenge requires a fundamental understanding of nanocrystal stability. In our research, we have worked towards such a goal by: (i) employing novel characterization techniques and developing syntheses that enable studies of the stability of nanocrystals,[2,3] (ii) investigating fundamental mechanisms of growth and stability of nanocrystals,[2,4] and (iii) integrating approaches from data science to accelerate our understanding and obtain a holistic view of nanocrystal stability.

Utilizing new characterization techniques can benefit our knowledge about nanocrystal stability. Taking Pt nanocrystals on high-surface-area carbon (Pt/C) as electrocatalysts as an example, we employed a novel approach to examine the real-time dissolution of Pt from nanocrystals. Specifically, an acid electrolyte is passed through a flow cell. The flow cell outlet is combined with an inductively-coupled plasma mass spectrometry (ICP-MS). The flow cell enables Pt/C nanocrystals to operate in their native environment, and the electrolyte flowing into the ICP-MS provides rich information about the stability of nanocrystals. In particular, we can distinguish at least three degradation mechanisms: (i) ionic dissolution, (ii) nanocluster detachment, and (iii) surface area changes.

Although novel characterization techniques can provide new insight into nanocrystal stability, synthesis limitations can also impact the investigation of nanocrystal stability with more conventional characterization techniques. Taking semiconductor magic-sized nanocrystals (MSNCs) as an example, we developed colloidal synthesis that enabled a more accessible investigation of MSNC growth and stability.[2,3] MSNCs are very small (< 2nm) nanocrystals with exceptional stability that grow in discrete steps jumping from one size to the next directly without forming intermediate crystallites.[2,5] The reason for the exceptional stability of "magic" size remains poorly understood. However, small sizes of MSNCs present significant challenges in their experimental investigation using conventional structural characterization techniques, e.g., they can be challenging to isolate and image via electron microscopy. We overcome these issues by developing a synthetic protocol for CdSe MSNCs using reactive organo-selenium precursors.[2,6] Our synthesis yields MSNCs up to larger length scales (> 2nm) that can be readily isolated and investigated further.

The synthesis of large MSNCs facilitates the investigation of their growth and stability. To determine the source of their stability, we applied a series of conventional optical, (UV-Visible spectroscopy, photoluminescence, photoluminescence excitation) and structural characterization techniques (X-ray diffraction, transmission electron microscopy). Our results suggest that MSNCs are a series of zinc-blende tetrahedral nanocrystals. To investigate their growth and stability, we isolated a particular size of MSNCs and heated them without precursors (conventional ripening). Our observations indicate that the MSNCs grow in the absence of precursors and exhibit size-dependent growth, and stability. These results from growth experiments were then explained by employing an atomistic model based on 3D and 2D-classical nucleation theory. Our model suggests that the exceptional stability of

MSNCs arises from their underlying tetrahedral shape and size-dependent growth barriers.[2]

Utilizing the insights obtained in the model above, we tried to modify the stability and growth trajectory of MSNCs. In particular, we explored three conditions (i) conventional ripening, (ii) ripening with sacrificial species (small MSNCs), and (iii) ripening with continuous addition of monomer.[4] Our results demonstrate that, even with remarkably distinct initial conditions, the growth trajectory of MSNCs synchronizes. More importantly, our findings indicated two important factors determining this growth synchronization: step and surface energy. By experimentally modulating the surface energy, we demonstrate methods to tune the stability and growth trajectory of MSNCs. The modification of surface energy also helped us to expand the size range of MSNCs further.[2,4]

The characterization techniques, syntheses, and growth experiments described above provide different approaches to gathering new insights about nanocrystal stability but are often resource intensive. Consequently, approaches are needed to accelerate our insights about nanocrystal stability. Here, we revert to our first system of Pt/C nanocatalysts. By combining a design-of-experiments-based (DoE) approach with online ICP-MS, we scan a vast parameter space to investigate the role of pre-treatment protocol on the stability of Pt/C nanocatalysts during electrocatalytic oxygen-reduction reaction. In particular, we employ a central composite design that generates a series of pre-treatment protocols for five different factors with five levels each. We then perform our experiments on the online ICP-MS described above. By running very few experiments (2 orders of magnitude < parameter space), our results identifies at least two key parameters (upper potential limit and potential depth) that affect dissolution. These findings help us to hypothesize key physical mechanisms that affect nanocrystal stability. Finally, we propose strategies to integrate such insights from statistical methods into more modern machine-learning-based algorithms (causal model). Thus, by employing novel characterization techniques, developing new syntheses, mechanistic investigation of growth, and integrating tools from data science, me and my colleagues work towards developing holistic understanding of nanocrystal stability.

(1) Kovalenko, M. V; Manna, L.; Cabot, A.; Hens, Z.; Talapin, D. V; Kagan, C. R.; Klimov, V. I.; Rogach, A. L.; Reiss, P.; Milliron, D. J.; Guyot-Sionnnest, P.; Konstantatos, G.; Parak, W. J.; Hyeon, T.; Korgel, B. A.; Murray, C. B.; Heiss, W. Prospects of Nanoscience with Nanocrystals. *ACS Nano* **2015**, *9*, 1012–1057.

(2) Mule, A. S.; Mazzotti, S.; Rossinelli, A. A.; Aellen, M.; Prins, P. T.; Van Der Bok, J. C.; Solari, S. F.; Glauser, Y. M.; Kumar, P. V.; Riedinger, A.; Norris, D. J. Unraveling the Growth Mechanism of Magic-Sized Semiconductor Nanocrystals. *J. Am. Chem. Soc.* **2021**, *143*, 2037–2048.

(3) Pun, A. B.; Mule, A. S.; Held, J. T.; Norris, D. J. Core/Shell Magic-Sized CdSe Nanocrystals. *Nano Lett.* **2021**, *21*, 7651–7658.

(4) Mazzotti, S.; Mule, A. S.; Pun, A. B.; Held, J. T.; Norris, D. J. Growth Synchronization and Size Control in Magic-Sized Semiconductor Nanocrystals. *ChemRxiv* **2022**.

(5) Pun, A. B.; Mazzotti, S.; Mule, A. S.; Norris, D. J. Understanding Discrete Growth in Semiconductor Nanocrystals: Nanoplatelets and Magic-Sized Clusters. *Acc. Chem. Res.* **2021**, *54*, 1545–1554.

(6) Riedinger, A.; Mule, A. S.; Knüsel, P. N.; Ott, F. D.; Rossinelli, A. A.; Norris, D. J. Identifying Reactive Organo-Selenium Precursors in the Synthesis of CdSe Nanoplatelets. *Chem. Commun.* **2018**, *54*, 11789–11792.

319e Macroscopic Evidence of the Liquidlike Nature of Nanoscale Polydimethylsiloxane Brushes

Xiaoxiao Zhao

We report macroscopic evidence of the liquidlike nature of surface-tethered poly(dimethylsiloxane) (PDMS) brushes by studying their adhesion to ice. Whereas ice permanently detaches from solid surfaces when subjected to sufficient shear, commonly referred to as the material's ice adhesion strength, adhered ice instead slides over PDMS brushes indefinitely. When additionally methylated, we observe Couette-like flow of the PDMS brushes between the ice and silicon surface. PDMS brush ice adhesion displays a shear-rate-dependent shear stress, rheological behavior reminiscent of liquids, and is affected by ice velocity, temperature, and brush thickness, following scaling laws akin to liquid PDMS films. This liquidlike nature allowsice to detach solely by self-weight, yielding an ice adhesion strength of 0.3 kPa, 1000 times less than a low surface energy, perfluorinated monolayer. The methylated PDMS brushes also display omniphobicity, repelling essentially all liquids with vanishingly small contact angle hysteresis. Methylation results in significantly higher contact angles than previously reported, nonmethylated brushes, especially for polar liquids of both high and low surface tension.

262b Liquid Crystalline Phase Behavior of Boron Nitride Nanotube Aqueous Dispersion

Joe Khoury, Eldar Khabushev, Lyndsey R. Scammell, Cheol Park, Angel Marti-Arbona, Matteo Pasquali

Boron nitride nanotubes (BNNTs), one-dimensional nanomaterials, possess a remarkable combination of mechanical, thermal, and optical properties that are incomparable with other materials. Their ability to self-assemble into long-range ordered domains, at high concentrations in different solvents, opens up exciting possibilities for the development of novel functional materials and devices with unprecedented properties and performance. However, obtaining highly concentrated and stable dispersions of BNNTs in different solvents remains a challenge due to the strong van der Waals attractions between adjacent nanotubes and the presence of large quantities of non-tubular impurities in the as-synthesized BNNTs. In this study, we announce the successful preparation of surfactant-assisted highly concentrated BNNT aqueous liquid crystalline dispersions, up to 18 mass% BNNTs. Polarized light microscopy (PLM) reveals the Schlieren texture of highly birefringent nematic liquid crystals. We identify the critical concentrations of BNNTs at which the phase transitions occur, which can aid in the design and optimization of BNNT-based materials. This study demonstrates that aqueous BNNT liquid crystals can be processed into transparent thin films. Scanning electron microscopy (SEM) and PLM reveal high BNNT alignment in the shearing direction, demonstrating control of macroscopic morphology and structure. The successful synthesis of highly concentrated BNNT aqueous liquid crystalline dispersions represents a significant advancement in the field and paves the way for the development of novel BNNT macroscopic materials for textiles, thermal management, electrical insulation, biomedical, and aerospace applications.

262c Rational Synthesis, Morphology, and Property Relationship of Non-Precious Bi-Metallic Hybrid Nanocomposites for Improving Oxygen Reduction Reactions

Mahshid Mokhtarnejad, Dibyendu Mukherjee, Bamin Khomami

Tailored metal oxide/metal-organic frameworks (MOFs) can readily be converted to hybrid nanocomposites (HNCs) comprised of transition metals interfaced with carbon (C) matrices. To that end, this class of nanocomposites (NCs) that exhibit superior conductivity and high porosity are promising and economically viable candidates as oxygen reduction reaction (ORR) electrocatalysts. Hence, optimizing the performance of this class of material by tailoring their composition, structure, size, and morphology through an efficient synthesis process has increasingly become an active area of research. In this study, we report the use of Laser Ablation Synthesis in Solution in Tandem with Galvanic Replacement Reaction (LASiS-GRR) technique as a facile, environmentally friendly, and rapid route for the synthesis and control of complex hierarchical bimetallic Zeolitic Imidazolate Framework (ZIF) structures, and their pyrolytic post-processing to yield metal oxide/MOF-based functional HNCs as highly active ORR electrocatalysts. The LASiS-GRR technique utilizes high-energy laser plasma reactions as the rate-controlling step for the initiation of solution-phase galvanic reactions. The one-pot, two-step process introduced in this study allows tailoring of the composition, structure, size, and morphology of the Co-based MOFs encapsulated in Zn-based porous MOF (Co/ZIF- 67@Zn/ZIF-8) crystals. Pyrolytic post-processing of these crystals leads to the creation of HNCs with superior functionality for ORR electrocatalysts for anion exchange membrane fuel cells (AEMFCs). Specifically, herein, for the first time, we demonstrate the use of the LASiS-GRR technique as a one-pot rational synthesis method for synthesizing Co/ZIF-67@Zn/ZIF-8 bi-MOFs that, upon post-pyrolytic treatments are converted into unique tailored HNCs of ZnO-interfaced ZIF-67 in carbonaceous matrices (ZnO/ZIF@C). Not only the effect of two distinct salt precursors on the particle size and morphology is investigated for the first time during LASiS, but also the superior performance and stability of these ZnO/ZIF@C HNCs as non-precious metal ORR electrocatalysts compared to commonly used Pt- and platinum-group metals (PGM)-based electrocatalysts in AEMFC applications is demonstrated.

262d Elucidating the Role of Surface Functionalization on the Photoluminescent Properties of Group IV Nanocrystals

Maharram Jabrayilov, Matthew Panthani

Presently, the globe is grappling with staggering climate changes stemming from the energy consumption. A potential way to prevail these challenges is to develop current technologies that optimize energy utilization with highest efficiency. Among many important energy consumption domains, microelectronics industry is one of the critical fields which uses substantial amount of energy for operation and also offsetting the heat dissipation from the components (such as memory chips). The reason for the consumption majorly comes from the conventional chips which transmit data with electrons, resulting in the resistive heating in the electronic parts. To address this issue, use of light as data transfer agent in chip-to-chip communications can substantially reduce the energy requirement. However, currently majority of these electronic parts are made up of silicon which does not have a capability of being emit light efficiently owing to the indirect bandgap nature. Thus, it is essential to develop novel materials to possess the ability to emit light efficiently for integrating it to the chip through advanced lithographic techniques

Herein, I will be presenting the colloidal synthesis and surface functionalization of Group IV nanocrystals (NCs), specifically Si and Ge and their alloys with Sn, for the optoelectronic applications. Colloidal synthesis of those materials offers great advantages such as low temperature synthesis, scalability and tunable optical properties. It also allows to synthesize nanocrystals with varying sizes and shapes. Furthermore, recently there is a lack of understanding how different surface termination impacts the electronic bandgap and optical properties in these nanomaterials. I will demonstrate the surface functionalization of the nanocrystals and their alloys with different organic and inorganic ligands and how it impacts the photoluminescent properties.

262e Deep Learning-Based Segmentation of Complex Microparticles in Scanning Electron Microscopy Images

Anastasiia Visheratina, Alexander Visheratin, Prashant Kumar, Michael Veksler, Nicholas Kotov

Nanoscale chirality is an actively growing research field spurred by the giant chiroptical activity, enantioselective biological activity, and asymmetric catalytic activity of chiral nanostructures.[1] Compared to chiral molecules, the handedness of chiral nano- and microstructures can be directly established via electron microscopy, which can be utilized for the automatic analysis of chiral nanostructures and prediction of their properties. However, chirality in complex materials may have multiple geometric forms. Its identification from electron microscopy images rather than optical measurements is fundamentally challenging because (1) image features differentiating left- and right-handed particles can be ambiguous, and (2) three-dimensional structure essential for chirality is 'flattened' into two-dimensional projections typical for electron microscopy images. Here, we show that deep learning algorithms can identify twisted bowtie-shaped microparticles[2] with nearly 100% accuracy and classify them as left- and right-handed with as high as 99% accuracy. Importantly, such accuracy was achieved with as few as 30 original electron microscopy images of bowties. Furthermore, after training on bowtie particles with complex nanostructured features, the model can recognize other chiral shapes with different geometries without re-training for their specific chiral geometry with 93% accuracy, indicating the learning abilities of the employed neural networks. These findings indicate that our deep learning algorithm trained on a practically feasible training set of experimental data enables automated analysis of microscopy data for the accelerated discovery of chiral particles and their complex systems for multiple applications.

Figure. Examples of segmentation of bowties using scanning electron microscopy images. Green color – right-handed particles, violet color – left-handed particles.

References

1) Cho, N.H., Guerrero-Martínez, A., Ma, J., Bals, S., Kotov, N.A., Liz-Marzán, L.M. and Nam, K.T., 2023. Bioinspired chiral inorganic nanomaterials. Nature Reviews Bioengineering, pp.1-19.

2) Kumar, P., Vo, T., Cha, M., Visheratina, A., Kim, J.Y., Xu, W., Schwartz, J., Simon, A., Katz, D., Nicu, V.P. and Marino, E., 2023. Photonically active bowtie nanoassemblies with chirality continuum. Nature, 615(7952), pp.418-424.

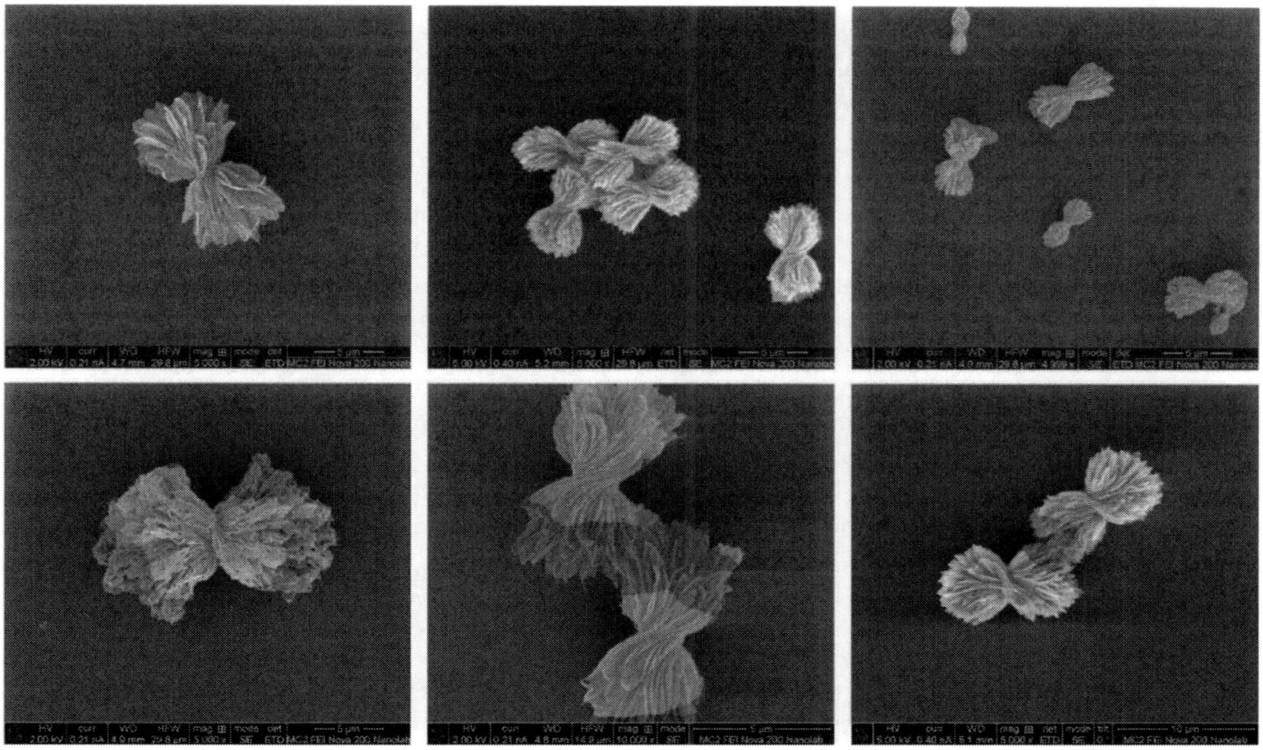

262f Minimizing the Formation of Polynuclear Palladium Hydroxo Complexes in the Mineralization of Barley Stripe Mosaic Virus

Che-Yu Chou, Akash J. Vaidya, Chamath Vindula Basnayake Pussepitiyalage, Shohreh Hemmati, Kevin Solomon, Sue Loesch-Fries, Michael T. Harris

Biotemplating has been developed as a green technique in the synthesis of uniform and controlled patterned nanostructures. The virion surface of barley stripe mosaic virus (BSMV), which is rich in functional biomolecules, is a desirable template for mineralization of nanomaterials. The hydrothermal synthesis has been demonstrated with the formation of uniform palladium coatings on BSMV without the addition of an extraneous reducing agent. Our recent focus is to investigate the formation of palladium polynuclear hydroxo complexes in the stock solution and during the mineralization process. The formation of palladium polynuclear hydroxo complexes can be attributed to the hydrolysis of palladium ions in aqueous media. As a result, the effect of palladium(II) hydrolysis on palladium polynuclear hydroxo species was studied. Hydrochloric acid was used to prevent the formation of palladium polynuclear complex in the Na_2PdCl_4 stock solutions. Dynamic light scattering (DLS) was used to monitor the formation of insoluble palladium polynuclear complex in Na_2PdCl_4 stock solutions. Transmission electron microscope (TEM) was used to image the synthesized Pd-BSMV nanorods. We have demonstrated that the Pd-BSMV nanorods synthesized from the HCl-added Na_2PdCl_4 stock solution resulted in the reduced formation of unwanted palladium polynuclear hydroxo complexes. In summary, this approach has intensified the hydrothermal synthesis technique and enhanced the quality of nanomaterials.

262g Atomically Precise Nanoclusters with Reversible Isomeric Transformation for Rotary Nanomotors

Zhaoxian Qin, Gao Li

Metal nanoclusters with well-identified crystal structures have drawn significant attention as emerging nanomaterials in fundamental research and application in fields including photoluminescence, catalysis, sensing, medicine, and biolabeling owing to their unique physical and chemical properties. Recently, thermal-stimuli responsive nanomaterials hold great promise in designing multifunctional intelligent devices for a wide range of applications. Despite recent advances in controllable synthesis and structure determination of atomically precise nanoclusters, the concept of reversible conformational isomerization for metal nanoclusters has not yet been explored. In this work, we report the discovery of a stimuli-responsive nanocluster that shows reversible conformational isomerism. The biicosahedral $[Au_{13}Ag_{12}(PPh_3)_{10}Cl_8]SbF_6$ nanoclusters composed of two icosahedral Au_7Ag_6 units by sharing one common Au vertex can produce two temperature-responsive conformational isomers with complete reversibility, which forms the basis of a rotary nanomotor driven by temperature. Differential scanning calorimetry analysis on the reversible isomeric transformation demonstrates that Gibbs free energy is the driving force for the transformation. This work offers a strategy for rational design of atomically precise nanomaterials, that the reversible stimuli-response behavior required for intelligent devices could be regulated via ligand tailoring and alloy engineering. The two temperature-driven, mutually convertible isomers of the nanoclusters open up an avenue to employ ultra-small nanoclusters (1 nm) for the design of thermal sensors and intelligent catalysts.

262h *Functionalization of Silicon with Tunable Bilayered Molecular System for Subsequent Photopatterning and Phosphorus-Based Dopant Attachment*

Patrick R. Raffaelle, Alexander Shestopalov

The ability to efficiently couple pattern-specific monolayers with dopants that demonstrate great structural variability is desired in the manufacturing of novel logic switching elements. In this study, we demonstrate a reliable protocol for the delivery of a phosphorus-based dopant onto functionalized silicon with precise spatial control at the microscale. This method relies on selective surface reactions which provide terminal functionalities that can be photochemically modified via UV-assisted contact printing between the modified surface and an elastomeric stamp inked with a molecular dopant. Spectroscopic measurements combined with high-resolution imaging microscopy was used to characterize the dopant attachment and patterning ability of this technique. Several notable features were observed in the resulting x-ray photoelectron spectroscopy (XPS) spectra that are indicative of the bonding behavior between dopant and monolayer including an increase in the atomic percentage of phosphorus on the silicon surface observed after photochemical printing. Scanning electron microscopy (SEM) analysis of the corresponding surface structures demonstrated high fidelity pattern transfer. The stability of the photopatterned surface in an ambient environment was also characterized using XPS and goniometry measurements, which demonstrated the high structural durability of the patterned surface across 72 hours of air exposure. Furthermore, an improved resistance to oxidation of the underlying silicon surface relative to unfunctionalized hydrogen-terminated silicon was also observed. Overall, this gentle approach to atomic precision surface processing has the potential to inform future development of next-generation devices in applications such as; quantum electronics, surface passivation techniques, nano-optics, and biosensors.

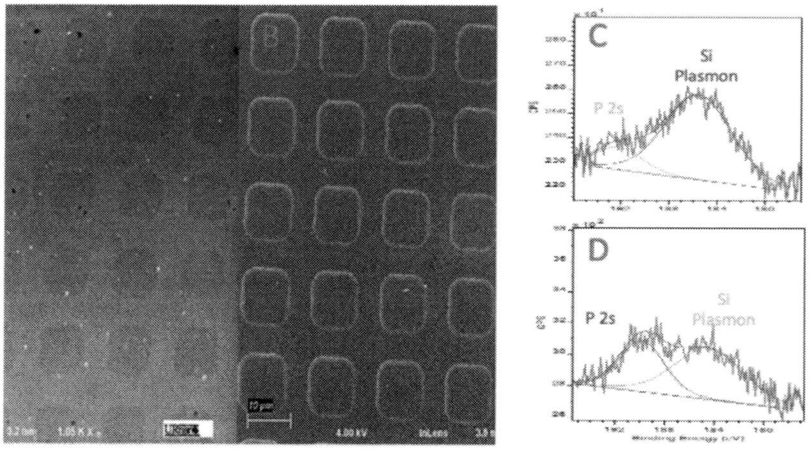

(Left) SEM images of: (A) silicon surface patterned with phosphorus-doped 8 μm squares, (B) corresponding SiO_2 stamp used to print dopant. (Right) XPS P 2s spectra measured from: (C) a bare silicon surface prior to stamping, (D) same surface following phosphorous-stamping. Post-stamping a new peak can be observed at 190.5 eV indicative of Si-P surface bonding.

426a DNA Patterning on Carbon Nanotube-Based Sensors Resolved By X-Ray Scattering Interferometry

Rebecca L. Pinals, Daniel J. Rosenberg, Francis J. Cunningham, Joshua Hubbard, Natalie S. Goh, Jeffrey Wang, Emily Hayman, Gregory Hura, Markita Landry

Single-walled carbon nanotubes (SWCNTs) with adsorbed nucleic acids have been applied in biomolecular sensing applications to investigate biological systems at requisite spatiotemporal scales. SWCNTs display tissue-transparent and photostable, yet sensitive, near-infrared fluorescence that render them uniquely suited for visualizing dynamic biological processes, including neurotransmitter signaling in the brain. Noncovalent adsorption of polymers such as DNA onto the SWCNT surface is implemented to disperse the otherwise hydrophobic SWCNTs in solution, while also providing a molecular recognition element that confers analyte specificity, giving rise to a nanosensor. Establishing rational design rules to inform development and refinement of such nanosensors requires an understanding of the discrete nanoscale architectures. To date, observations have been restricted to computational models or experiments completed under dehydrated states, despite expected morphological differences in solution and ultimate application within aqueous biological environments.

In this work, we demonstrate a direct mode of measuring in-solution nucleic acid geometries on SWCNTs via high-throughput X-ray scattering interferometry (XSI). This technique leverages the interference pattern produced by ordered AuNP tags conjugated to the DNA on the SWCNT surface. We employ XSI to quantify distinct surface-adsorbed morphologies for two different DNA oligomer lengths, conformational changes as a function of ionic strength, and the mechanism of dopamine sensing for a previously established DNA-SWCNT nanosensor technology, with corresponding *ab initio* modeling for visualization. We find that the shorter ssDNA adopts a highly ordered structure of alternatingly oriented rings stacked along the SWCNT axis, in comparison to the longer ssDNA that wraps helically. For the shorter ssDNA, the presence of the target analyte dopamine leads to simultaneous axial elongation and radial tacking of the ssDNA closer to the SWCNT surface, distinct from conformational changes elicited in the presence of high ionic strength. Application of XSI to probe solution-phase morphologies of nanoparticle-based tools will elucidate the mechanism of sensor function and guide future design strategies.

426b Programming DNA Self-Assembly for Molecular Robotics and Responsive Materials

Alexander Marras

Synthesizing biomaterials using nature's building blocks enables engineers to construct highly programmable devices with unprecedented control over physical and chemical properties, providing a basis for next generation smart materials. We use structural DNA nanotechnology to design and build controllable nanoscale mechanisms and machines. Leveraging the unique structural properties of DNA, we created mechanical joints and multi-joint mechanisms with defined 2D and 3D motion. We then used these mechanisms to demonstrate methods for mechanical actuation, demonstrating conformational control of dynamic nanostructures using DNA triggers or responding to environmental conditions. I will discuss how our mechanisms can be used for ultra-sensitive detection of chemical, physical, and biological cues and how they can be assembled to into dynamic multi-scale materials.

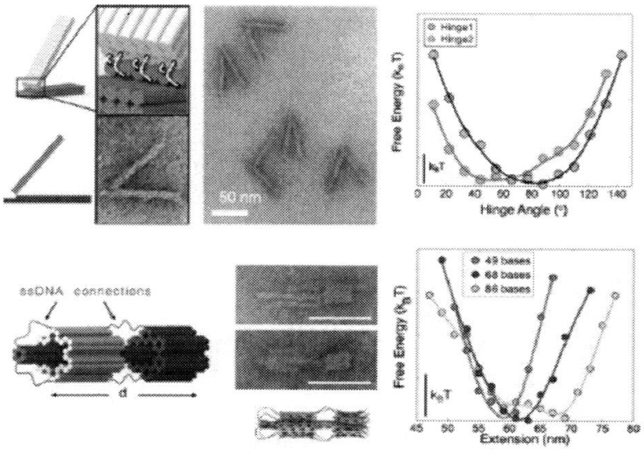

Tunable DNA-based mechanical joints

Multi-joint mechanisms

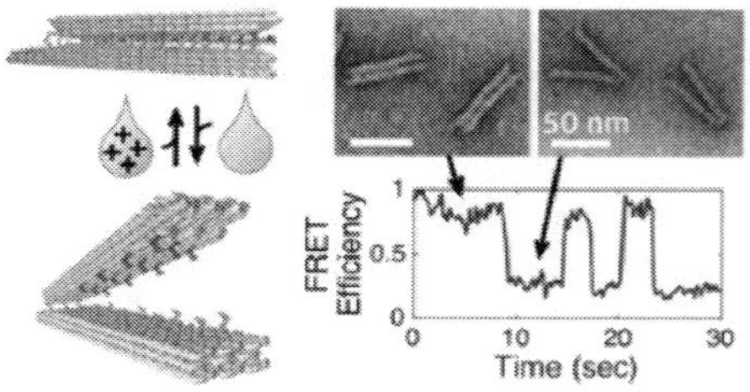

Rapid actuation studied with single-molecule FRET and TEM

426c Enhancing Clot Formation and Mechanical Strength with Graphene / Peg Hemostatic Nanoplates

Tannaz Farrokhi, Manos Gkikas

For those under the age of 44, as well as those serving in the armed forces, trauma (blast or blunt) is the main cause of mortality. A number of methods have been proposed to effectively halt bleeding from external wounds. Nonetheless, dealing with internal bleeding is complicated. When an injury occurs, hemostasis (the natural clotting mechanism) begins to stop the bleeding and start the recovery process. However, *many patients suffering from lethal internal injuries decease from severe blood loss while being transferred to the hospital* and before even reaching the emergency room where they can be treated by surgeons. An effective approach would be the development of synthetic hemostatic formulations that could be administered by paramedics during a patient's transport to the hospital and reduce blood loss, by augmenting the natural clotting mechanism and leading to faster formation of clots.

In the natural clotting mechanism, fibrinogen binds to activated platelets and crosslinks those. Most studies have thus been focused on ***enhancing hemostasis by interacting activating platelets integrins (GPIIb/IIIa)*** with***fibrinogen-derived peptides*** *(either from the α- or the γ-chain)* that are ***conjugated to either liposomes or polymer-based nanoparticles***. A second major concern expressed by surgeons regards the stability of clots post formation, especially when are under mechanical stress and blood flow. ***Graphene oxide (GO)*** has been used in a variety of biomedical applications due to its *large surface area and mechanical properties*. Aiming to provide ***dual hemostatic ability*** upon interaction with activated platelets as well as to ***augment the fibrin clot strength and durability upon integration***, we are developing a ~300-500 nm ***nanographene-based hemostat***. The nanoGO/PEG material is ***functionalized with two peptides from both the α-chain and the γ-chain of fibrinogen***, offering interaction with activated platelets *at two binding sites*. The material is evaluated *in vitro* with human blood platelets activated with ADP and collagen for platelet plug enhancement / reduction in bleeding time (clotting acceleration) over controls, and *ex vivo* for mechanical enhancement of blood clots (tensile stretching); next, it will be evaluated *in vivo* with a lethal injury animal model after intravenous administration.

426d Nanostructured, Biocompatible Mxene-Based Hybrid Membranes Fabricated Using Vacuum-Assisted Layer-By-Layer Assembly

Gelareh Rezvan, Farivash Gholamirad, Mary Walden, Monirosadat Sadati, Nader Taheri-Qazvini

This study presents the development of a flexible, self-assembling hybrid membrane composed of two-dimensional titanium carbide MXene nanoflakes and gelatin-stabilized amorphous calcium phosphate (Gel-ACP) nanoparticles. We hypothesized that the controlled integration of Gel-ACP nanoparticles and MXene nanoflakes in a nanostructured hybrid membrane would result in enhanced mechanical strength, electrical conductivity, and biocompatibility, making it suitable for biosensing and tissue engineering applications. Two-dimensional $Ti_3C_2T_x$ MXene nanoflakes were synthesized by selectively etching Al layers from the Ti_3AlC_2 MAX phase. Gel-ACP nanoparticles with an average size of 50 nm were synthesized through controlled precipitation of calcium phosphate in the presence of gelatin. The hybrid membranes were fabricated using vacuum-assisted layer-by-layer assembly of Gel-ACP nanoparticles and MXene nanoflakes. Nanostructural analysis was performed using electron microscopy and small-angle and wide-angle X-ray scattering. Furthermore, mechanical properties were assessed through a newly designed apparatus for measuring membrane burst pressure. Electrical conductivity and biocompatibility were also evaluated to determine the membrane's potential for biomedical applications. The hybrid membranes exhibited exceptional mechanical properties due to the strong interfacial interactions between the Gel-ACP and MXene layers. The membrane's flexibility allowed it to endure deformation while maintaining structural integrity. The MXene nanoflakes provided electrical conductivity, an advantageous feature for applications requiring electrical stimulation. Additionally, the membrane displayed biocompatibility, excellent adhesion to bone-like structures, and long-term stability in buffer solutions. The innovative design of the MXene/Gel-ACP membrane showcases its potential for diverse applications, including guided bone regeneration, benefiting from the membrane's bioactivity and mechanical support. This study provides a foundation for future research into the MXene-based hybrid membrane's potential in tissue engineering and diagnostics, further advancing the development of nanostructured biomimetic and biohybrid materials.

30t Next generation nano-diagnostic devices: Graphene interaction with skin cancer cells

Kartikey Sharma

In this study, we exploit the remarkable properties of graphene, a versatile two-dimensional nanomaterial, to discern and differentiate between cancerous and normal tissues through subtle modifications in graphene's characteristics. The inherent sensitivity of graphene's zero-band structure to doping phenomena, stemming from interactions with surrounding materials, forms the foundation of our approach. Leveraging this phenomenon, we investigate the distinct Raman spectral features of graphene, specifically the 2D and G bands, which exhibit distinctive blue or red shifts contingent upon the nature of doping. Through a meticulously designed 3D-printed platform, graphene is meticulously interfaced with basal cell carcinoma (BCC), squamous cell carcinoma (SCC), melanoma, and normal skin tissues on glass slides. Notably, our investigation of graphene's Raman spectra reveals a profound shift in the 2D band for normal versus cancerous tissues, underscoring its potential as a discriminatory diagnostic tool. This assertion is substantiated by consistent findings across 4 BCC, 4 SCC. Our work holds profound implications for non-invasive and precise cancer diagnosis, heralding a new era in early detection and personalized therapeutic interventions.

426f Scalable Nanofabrication of Biomimetic Broadband Antireflection Coatings

Fan Gao, Jiachen Wu, Peng Jiang

Millions of years before people began to fabricate functional nanostructures, biological systems were using nanometer-scale architectures to produce unique functionalities. Some nocturnal moths use hexagonal arrays of subwavelength nipples as antireflection (AR) coatings to reduce reflection from their compound eyes for the purpose of night camouflage. Similar periodic arrays of nanopillars have also been observed on the wings of cicada to render superhydrophobic surfaces for self-cleaning functionality. Inspired by these natural nanostructures, we have developed a simple and scalable templating technique for fabricating self-cleaning, broadband AR coatings on a large variety of technologically important substrates, such as crystalline Si, sapphire, GaAs, and glass. The technique is based on two colloidal self-assembly technologies (i.e., spin coating and Langmuir-Blodgett coating) that combines the simplicity and cost benefits of bottom-up self-assembly with the scalability and compatibility of standard top-down microfabrication. The resulting subwavelength-structured AR coatings exhibit improved broadband antireflection properties than traditional multilayer dielectric coatings, and show superhydrophobic surface states, promising for applications ranging from highly efficient solar cells and photodetectors to self-cleaning flat-panel displays.

426g Protein Aggregation on Metal-Oxide Nanozymes Governs Activity and Cellular Uptake

Robert Nißler, Lena Dennebouy, Alexander Gogos, Michael A. Pais, Anna L. Neuer, Monika Zimmermann, Ioana Lese, Inge Herrmann

Catalytically active inorganic nanomaterials offer promising prospects for biomedicine. Inorganic enzyme mimicking nanoparticles, termed nanozymes, have enabled fascinating applications, including attenuation of ischemia/reperfusion injury, and improved wound healing through radical scavenging and cell protection. Similarly, metal oxides can also be employed to generate radicals and induce cell death, for example in photodynamic or radiotherapy scenarios. When applying these nano-enzymes into a biological system, proteins rapidly physisorb on the surface of the metal oxide nanozymes, thus altering their physiochemical properties. We quantified and characterized the fraction of adsorbed proteins by proteomic analysis and linked the observed proteins to the nanoparticle fate *in vivo*. Opsonization of nanozymes led to an enhanced uptake by macrophages following topical application of the metal oxide nanozyme suspensions, and particle largely remained at the site of application (>99%). The enzyme mimicking activity was affected by the presence of proteins to a variable extent, suggesting that shielding of the nanoparticle surface by proteins needs to be taken into account in the development of catalytically active nanomaterials. Finally, we present potential strategies for the prevention and recovery of the catalytical activity based on metal oxide nanoparticle surface engineering. In summary, we investigated the influence of physiosorbed proteins on radical scavenging and radical generation based on a selection of catalytically active metal oxide nanoparticles, combining state-of-the-art proteomics and nanoanalytics, providing design criteria for high performance nanocatalysts exhibiting activity in complex biological environments.

426h Direct Observation of Heterogeneous Distribution of Enzymes in Metal-Organic Frameworks and Its Effects in Propelling Nanomotors

Yu Liu, Jun Ge

Natural biomineralization processes have inspired the *in situ* co-precipitation synthesis of protein-metal-organic frameworks (pMOFs) with potent competitiveness in the field of catalysis, imaging, sensing, and drug delivery. Understanding of crystallization process and biomolecule behavior of these hybrid composites helps guide the design of multifunctional pMOFs with enhanced activity and stability. While efforts have been made to investigate the nucleation and growth mechanisms of pMOFs, the specific protein distribution pattern and its role in biocatalytic functions have not been fully elaborated.

Here, we reconstructed the three-dimensional protein distribution inside pMOF composites on the molecular level by super-resolution optical microscopy and clustering analysis. Via detecting, correcting, and Gaussian fitting of signals from individually excited fluorophores, the specific localization of each detection can be acquired to overcome the resolution limit of traditional fluorescence microscopy. With clustering analysis methods, protein clusters were identified inside the porous solid particles, revealing the heterogeneity underlying the apparent symmetry. Detections centralized mostly at the center of the particles, indicating the important role protein played during the nucleation of pMOF composites. Moreover, for individual particles, the co-precipitation process promoted the scattered localization of protein clusters which generated a structural and functional asymmetry.

These results inspired us to investigate the potential of pMOFs themselves as biocatalytic nanomotors that required asymmetry for active propulsion at low-Reynold's number, a ubiquitous environment for nanoparticles in many biocatalytic and biomedical applications. Catalase@ZIF-8 (CAT@ZIF-8) was synthesized as a model and exhibited both enhanced diffusion in uniform substrate solution and positive chemotaxis in microfluidic chips with a gradient concentration. We then proposed a simple model to verify numerically that clustered protein distribution generated stronger self-diffusiophoretic propulsion than complete spatial random distributions. Different proteins exhibited similar clustered behavior, and enzyme cascades can be easily immobilized as well to increase the energy utilization efficiency of fuels and amplify the propulsion power.

In summary, the combination of super-resolution optical microscopy and clustering analysis provided a potent method to reconstruct specific protein distribution in pMOF composites. Collectively, protein clusters were found to be stochastically centralized mainly at the center of particles, which originated from protein-dependent nucleation and crystallization process. Individually, co-precipitation promoted the scattered localization of protein clusters which generated a structural and functional asymmetry leading to the active propulsion of biocatalytic pMOFs. Taken together, these results offered new insights into comprehending the co-precipitation process from a fresh perspective.

158a One-Step Fluorine-Free Synthesis of Delaminated, OH-Terminated Ti_3C_2: High Photocatalytic NO_x Storage Selectivity Enabled By Coupling TiO_2 and Ti_3C_2-OH

Ahmed Al Mayyahi, Swagotom Sarker, Brian M. Everhart, Xiaoqing He, Placidus Amama

Difficulties associated with MXene fabrication typified by environmentally hazardous fluorine-based etching process followed by delamination and fluorine reduction steps create obstacles to its expansion and application in photocatalytic air purification. Herein, we demonstrate a one-step hydrothermal process that involves alkali (NaOH)-assisted aluminum etching in the presence of a delaminating agent (hydrazine) for the synthesis of delaminated, OH-terminated MXene (Ti_3C_2-OH). The process does not require fluorine-containing compounds and produces Ti_3C_2-OH sheets without post-synthesis treatment. Coupling Ti_3C_2-OH with TiO_2 shows a robust synergistic effect as a co-catalyst in the photocatalytic oxidation of NO_x. Excellent NO_x storage selectivity (91%) and a positive $DeNO_x$ index (+ 0.30) were achieved at NO conversion of 54%. The photocatalytic activity of the hybrid shows high sensitivity to the surface chemistry of the co-catalyst as Ti_3C_2-OH outperforms delaminated, F-terminated Ti_3C_2 (fabricated via traditional process), which shows low NO_x storage selectivity (65%) and a negative $DeNO_x$ index (-0.05) at NO conversion of 47%. These findings will benefit ongoing efforts to develop effective alternative methods for MXene synthesis and provide a pathway for the rational design of efficient NO_x abatement photocatalysts.

158b Controlled Synthesis of 2D Transition Metal Dichalcogenides for Electronic Nanosensors

Meng-Qiang Zhao

Modern electronics and optoelectronics require large-scale growth of semiconducting thin films, which would benefit the development of high-performance ultrathin and flexible devices. Two-dimensional (2D) transition metal dichalcogenides (TMDs) provide ideal semiconducting materials with attractive electronic and photonic properties. Towards to route to their practical applications, the controlled growth of 2D TMDs with well-designed composition, layer number, size, and structure is prerequisite. In this talk, I will present our recent progress on the controlled growth of 2D TMDs by chemical vapor deposition (CVD). Bilayer TMDs with a 100% selectivity, a variety of TMD-based heterostructures, and large-area continuous monolayer TMD films are all successfully obtained. Beyond TMDs, ongoing research activities also focus on the controlled synthesis of 2D transition metal carbides, known as MXenes. An universal functionalization method based on hexagonal boron nitride (hBN) interlayers has been developed to promote the application of 2D semiconductors in electronic nanosensors for the detection of DNAs and organic pollutant molecules.

158d Chemical Synthesis and Physical Properties of Graphimine - a New 2D High-Performance Polymer

Kätchen K. Lachmayr, David C. McLeod, Eric D. Wetzel, Steven Lustig

Graphimine is a member of a new class of two-dimensional polymers composed of hexa-substituted benzene nodes within a two-dimensional covalent organic framework (2D COF). Drawing inspiration from robust materials like graphene, graphimine comprises a rigid, aromatic network with short linkages between benzene nodes, thus having the potential to revolutionize the current standards for lightweight, high-performance materials. Graphimine is synthesized in solution from a novel hexa-substituted monomer, 2,4,6-triaminobenzene-1,3,5-tricarbaldehyde (TABTCA). TABTCA was derived from a high yielding, three-step, synthetic pathway, utilizing a substitution reaction with anhydrous ammonia and 2,4,6-tribromobenzene-1,3,5-tricarbaldehyde. Nuclear magnetic resonance spectroscopy was used to successfully confirm its chemical structure, and mass spectrometry verified its molecular weight. Polymerization of TABTCA into graphimine is conducted using reversible, acid-catalyzed polycondensation, that results in imine linkage formation, and structural re-organization via transimination to achieve the 2D COF. Graphimine is analyzed with FTIR, to determine conversion of aldehyde and amine functionalities to imine-linkages; PXRD, to analyze the resulting crystalline structure; and HRTEM, to examine interlayer stacking structure. Furthermore, graphimine has been oxidized to graphamide, which is a dense 2D polyaramid. Graphimine and graphamide are fabricated into films using techniques such as drop casting, spin-coating, and doctor blading, for applications that require extended sheets or films. These results provide evidence for the first successful production of a highly compacted imine- and amide- linked two-dimensional polymers and hold great potential for development as a new low-weight, high-strength, performance materials.

158e The Effects of Ionic Strength on Rheological Properties and Microstructural Changes in Aqueous Mxene ($Ti_3C_2T_x$) Dispersions

Mackenzie Woods, Fatima Hamade, Majid Beidaghi, Virginia Davis

This work examines the effects of salt addition and concentration on the rheological properties and phase transitions of aqueous MXene ($Ti_3C_2T_x$) dispersions. MXenes are a relatively new family of two-dimensional (2D) nanomaterials comprised of transition metal carbides and nitrides. Their exceptional electrical, electrochemical, and thermal properties have made them a desirable candidate for fluid phase processing of advanced devices. Understanding nanomaterial rheology and phase behavior is of practical relevance for the improvement of fluid phase manufacturing processes such as direct ink writing. Dispersions of large (~3 μm) MXenes and a bimodal mixture of large and small (~0.3 μm) MXenes were investigated over a range of concentrations. Sodium chloride (NaCl) was added to specific concentrations to investigate the effects of increasing ionic strength on dispersion behavior and properties. Steady shear and small amplitude oscillatory shear rheology were used to understand the change in viscoelastic properties and flow behavior of the dispersions as a function of MXene concentration and ionic strength. Cross-polarized optical microscopy (POM) was utilized to observe changes in microstructure and dispersion morphology as MXene and salt concentrations were increased. Without the addition of NaCl, the large MXene dispersions readily arranged into extended aligned structures, a feature that typically advances fluid phase processability. Bimodal MXene dispersions also formed extended aligned structures, though a higher concentration was needed to achieve similar results to the large MXene dispersions. POM has shown that birefringence is observable in both large MXenes and bimodal MXenes, suggesting that the liquid crystalline phase can be achieved in these additive-free dispersions. The addition of a low concentration of NaCl resulted in an accelerated transition to a gel-like phase for large MXene dispersions, with a similar phase behavior change in bimodal MXene dispersions. Further increasing NaCl concentration caused severe aggregation and discontinuity of microstructure in both large and bimodal MXene dispersions. These results provided a deeper understanding of a new 2D nanomaterial/salt system and the parameters affecting dispersion tunability. The insights achieved from this work provide a foundation for further optimization of MXene printing inks and devices.

158f Programming Self-Assembly of Plasmonic Nanomaterials with Sequence-Controlled DNA Polymers

Yifeng Shi, Grigory Tikhomirov

DNA origami – a DNA-based polymer – has been generally considered as a powerful tool to achieve the self-assembly of inorganic plasmonic nanocrystals due to its high programmability and addressability. However, the assembly strategies developed so far are mostly limited to two sub-categories: i) non-finite lattice whose final dimension and shape cannot be controlled and ii) small-scale finite structures with typically only 2-10 nanocrystals. Moreover, in both types of plasmonic nanocrystal-DNA origami hybridization systems mentioned above, the plasmonic nanocrystals are only partially connected to the origami, which limits the number of addressable functionalization sites associated with each nanocrystal. Here we report a new strategy for programmable assembly of gold nanorods that can be assembled into a finite lattice with high tunability in their final configuration. The assembly of the nanorods is achieved using a simple rectangular DNA origami. The nanorods were first "wrapped" by the origami, and these wrapped nanorods were then assembled into finite lattice with desired 2D/3D configuration based on the pre-designed extension sequence position on the origami rectangles. By adjusting the dimension of the origami, or the length and sequence of the extensions, we could correspondingly adjust the size of the assembly unit, their interspacing, and the assembly pattern. Since almost the whole surface area of the nanorod is covered by the origami, these nanorods becomes more programmable due to a larger area of possible functionalization sites. Moreover, since the electric field enhancement is proportional to the particle size, the larger nanorods applied in the current study could greatly enhance their hotspot effect upon assembly. The well-defined configuration and enhanced hot-spot effect makes this nanorod assembly promising in a variety of fields including Surface Enhanced Ramen Spectroscopy (SERS), plasmonic photocatalysis, and plasmonic-enhanced luminescence.

352a Tracking Dendritic Cell Vaccination Using Magnetic Particle Imaging

Bo Yu, Ginger Moore, Alexander Armstrong, N'quel Bonner, Sitong Liu, Ambar Velazquez Albino, Duane A. Mitchell, Carlos Rinaldi-Ramos

Dendritic cell (DC) based vaccines have proven to be a safe therapeutic approach, although with inconsistent clinical results. DCs must migrate to lymphoid organs to be effective and the magnitude of anti-tumor response is expected to be correlated to DC migration to lymph nodes (LNs). It's reported that a small fraction of administered DCs migrate to LNs. Therefore, non-invasive methods to sensitively monitor and quantify migration of adoptive DC therapies would be tremendously helpful in evaluating immune response. Although techniques such as magnetic resonance imaging (MRI), single-photon emission computed tomography (SPECT), and positron emission tomography (PET) have previously been used to track DCs, these techniques are limited by low cell detection sensitivity in the range of 10^4 cells or more.

Magnetic particle imaging (MPI) is a novel molecular imaging technology that can sensitively and quantitatively detect iron oxide tracers *in vivo*, with higher cell detection sensitivity compared to other imaging approaches. Cell detection sensitivity in MPI is determined by two factors: the properties of the tracer used to label cells and the extent of tracer uptake per cell. Other than detection sensitivity, to be able to differentiate signals from LNs and injection sites, resolution and imaging conditions are crucial factors in tracking DC migration *in vivo*.

In this study, cellular uptake was evaluated for MPI tailored tracers (RL-1) coated with poly(maleic anhydride-alt-1-octadecene) (PMAO). The MPI maximum signal intensity for the tracer was 101 $mgFe^{-1}$. The upper labeling limit for marrow-derived dendritic cells (BMDCs) was determined to be ~ 50 pgFe/cell by screening SPION concentration and incubation time. Cell viability, functionality, and surface biomarker expression were not affected by tracer labeling. Confocal images suggested particles are in vesicle-like structures inside cells. *In vitro* tracer fate studies suggested particles are not degraded or exocytosed after 3 days of culture. Cell dilution studies demonstrated detection of as little as 10^3 cells with MPI using optimized labeling conditions. An *in vivo* BMDC tracking study was performed using high sensitivity and high-sensitivity/high-resolution scan modes and different administration routes (foot pad subcutaneous injection and hind flank intradermal injection). Administered DC numbers were quantifiable using *in vivo* day 0 MPI measurements. BMDC migration quantified by *ex vivo* lymph node MPI measurements ranged from 3000 to 5000 cells, however, signals in LNs *in vivo* were not detectable because the strong signal from the injection site masked the weak LN signal. An *ex vivo* phantom study was performed to evaluate signal differentiation and quantification accuracy as a function of separation distances, administered doses, and imaging settings. Then an *in vivo*BMDC tracking experiment using advanced imaging settings with partial filed of view and full field of view was performed, in vivo BMDC migration was trackable with partial field of view imaging. The correlation of in vivo LN MPI signal and ex vivo LN MPI signal was evaluated. BMDC migration efficiency was compared with and without GM-CSF. A survival study will be performed to evaluate correlation of LN MPI signal and DC vaccination therapeutic response. Together, these studies demonstrate the potential of MPI as tool for quantitative tracking of DC migration in DC vaccination studies.

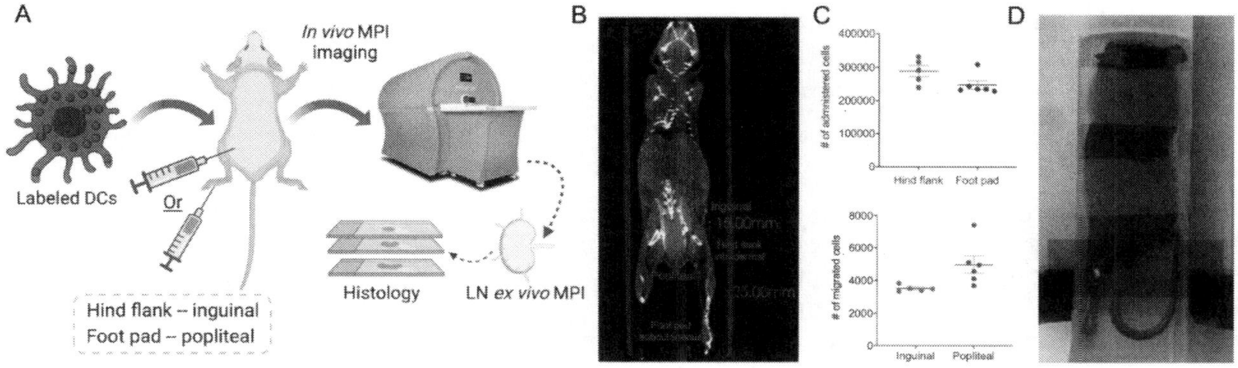
A
Labeled DCs
Or
In vivo MPI imaging
Hind flank -- inguinal
Foot pad -- popliteal
Histology
LN ex vivo MPI
B
C
# of administered cells
400000
300000
200000
100000
0
Hind flank
Foot pad
# of migrated cells
8000
6000
4000
2000
0
Inguinal
Popliteal
D

352b Facile and Quantitative Detection of Nucleic Acid Markers Via Conformational Change of Gold Nanoparticle Assemblies

Yuto Sahara, Noriko Nakamura, Seiichi Ohta

Introduction

Nucleic acid is a promising biomarker for clinical diagnostics. In recent years, the relationship between various types of nucleic acids, such as messenger RNA (mRNA), micro RNA (miRNA), cell free DNA (cfDNA), and diseases has been revealed. In addition, since nucleic acids with short length, such as miRNAs, are contained in various body fluid including saliva and blood, they can be easily collected from the body. These characteristics makes nucleic acids as attractive target for diagnostics.

Polymerase Chain Reaction (PCR) is one of the most widely used nucleic acid detection methods. However, although PCR can quantify target nucleic acids with high sensitivity, it requires complicated temperature operations. As a more facile method, use of surface plasmon coupling of gold nanoparticles (Au NPs) has attracted attentions. Previous studies reported that by modifying AuNP surface with single stranded DNA that is complementary to target nucleic acid markers, the existence of target markers triggers their aggregation by inter-particle crosslinking via double helix formation, resulting in the color change from red to purple by plasmon coupling [1]. This strategy enables simple and rapid detection of nucleic acid markers by just mixing the sample. However, since large scale aggregation of AuNPs induces their sedimentation, absolute quantification of target concentration via optical absorption is difficult. Therefore, the development of a rapid, facile, and quantitative nucleic acid detection method is expected.

In this study, to achieve plasmon coupling between AuNPs without causing sedimentation, we focused on DNA-mediated assembly of gold nanoparticles into superstructures. We previously reported that by using complementary DNA as a linker, single stranded DNA-modified AuNPs can be assembled into core-satellite type superstructure. The assembled superstructure can then be transformed into different conformation via reconstruction of DNA double helix with another specific DNA, called as toehold-mediated strand displacement [2]. By using this assembly technology, in this study, we aimed to design AuNP assemblies so that they show purple color via close inter particle distance without causing sedimentation at the initial state, whereas they dissociate into individual AuNPs in response to specific nucleic acid target via toe hold displacement, resulting in the color shift into red. It is expected that by changing the strategy from 'dispersion to large aggregation' to 'controlled aggregation to dispersion', facile and quantitative nucleic acid detection can be achieved.

Experimental

AuNPs with sizes of 8, 15, 40, and 90 nm were synthesized by the reduction of tetrachloro gold(III) acid with citric acid, tannic acid and hydroquinone. The surface of the obtained AuNPs were modified with thiol-terminated single stranded DNA. DNA-modified AuNPs were further assembled into core-satellite superstructure by mixing with complementary DNA linkers, which was designed to dissociate in response to target nucleic acid sequence via toehold-mediated strand displacement. Structure and optical property of AuNP assembly before and after the reaction with target nucleic acids were evaluated by TEM and UV-vis spectroscopy, respectively.

Results and Discussion

Core-satellite superstructure was assembled using AuNPs with different size combinations: 15nm core-8nm satellite, 40nm core-15nm satellite, and 90nm core-40nm satellite. Successful assembly was confirmed by TEM observation. Red shift of the absorption wavelength induced by the superstructure formation was observed for all size combination of building block AuNPs. Comparing between different sizes, use of larger AuNPs resulted in longer shift in absorption wavelength, which was consistent with the theory of surface plasmon coupling. Since 90nm core-40nm satellite assemblies possibly cause gravitational sedimentation during the following assay due to their large size, 40nm core-15nm satellite assemblies were used as a representative formulation hereafter.

To test the dissociation of AuNP assembly in response to specific nucleic acid target, we first prepared the model target DNA with different length and complementary position. By the addition of the model target DNA, AuNP assembly was dissociated into individual building block particles, regardless of the types of model target DNA. Furthermore, this conformational change induced distinct blue shift of the absorption wavelength up to 11 nm, suggesting the possibility for colorimetric target nucleic acid quantification. Whereas final structure and consequent absorption wavelength of AuNP assembly was same between model target DNA with different length and complementary position, this difference significantly altered the dissociation kinetics. In the case of short target DNA (36 base), absorbance change at 525 nm, which corresponds to the degree of AuNP dissociation, rapidly occurred within 40 min. The time course also followed the second order reaction kinetics, estimated by the theoretical model for toehold-mediated strand displacement of DNA. On the other hand, if the model DNA became longer or the complementary position in the model DNA became far from the terminal end, the dissociation reaction kinetics became slower and apart from the theoretical line. This delay would be attributed to the steric hinderance of model DNA when approaching to the linker DNA strand surrounded by many AuNP satellites. From these results, we selected miRNA, which has relatively short length (ca. 22 base) compared with other nucleic acid marker, as a representative marker for the following quantification assay.

To investigate if the constructed AuNP assembly can be used for the quantification of actual nucleic acid biomarker for diagnostics, cancer marker miR-21 was selected as a target. miR-21-responsive AuNP assemblies were designed similarly with the above experiments, and then exposed to miR-21 with various concentrations. As a result, dissociation of AuNPs occurred in dose-dependent manner, faster with increasing miR-21 concentration. Reaction rate at the initial 10 min showed linear relationship with miR-21 concentration, demonstrating that our AuNP assembly system can achieve colorimetric, absolute quantification of cancer miRNA marker by just mixing without causing aggregation-induced sedimentation.

Conclusions

We developed AuNP assembly-based assay system for facile and quantitative detection of nucleic acid biomarkers. AuNP assembly dissociated into individual particles in response to target nucleic acids, leading to the shit in absorption wavelength via change in surface plasmon coupling. The kinetics of AuNP dissociation was dependent on the length and complementary position of target nucleic acid, due to steric hinderance. The absolute quantification of cancer biomarker using the developed assay was demonstrated using miR-21 as a target.

Acknowledgement

This work was supported by JST PRESTO.

References

1. A. Reynolds, C. A. Mirkin, and R. L. Letsinger: *J. Am. Chem. Soc.*, 2000, 122, 3795.
2. S. Ohta, D. Glancy, and W. C. W. Chan: *Science*, 2016, 351, 841.

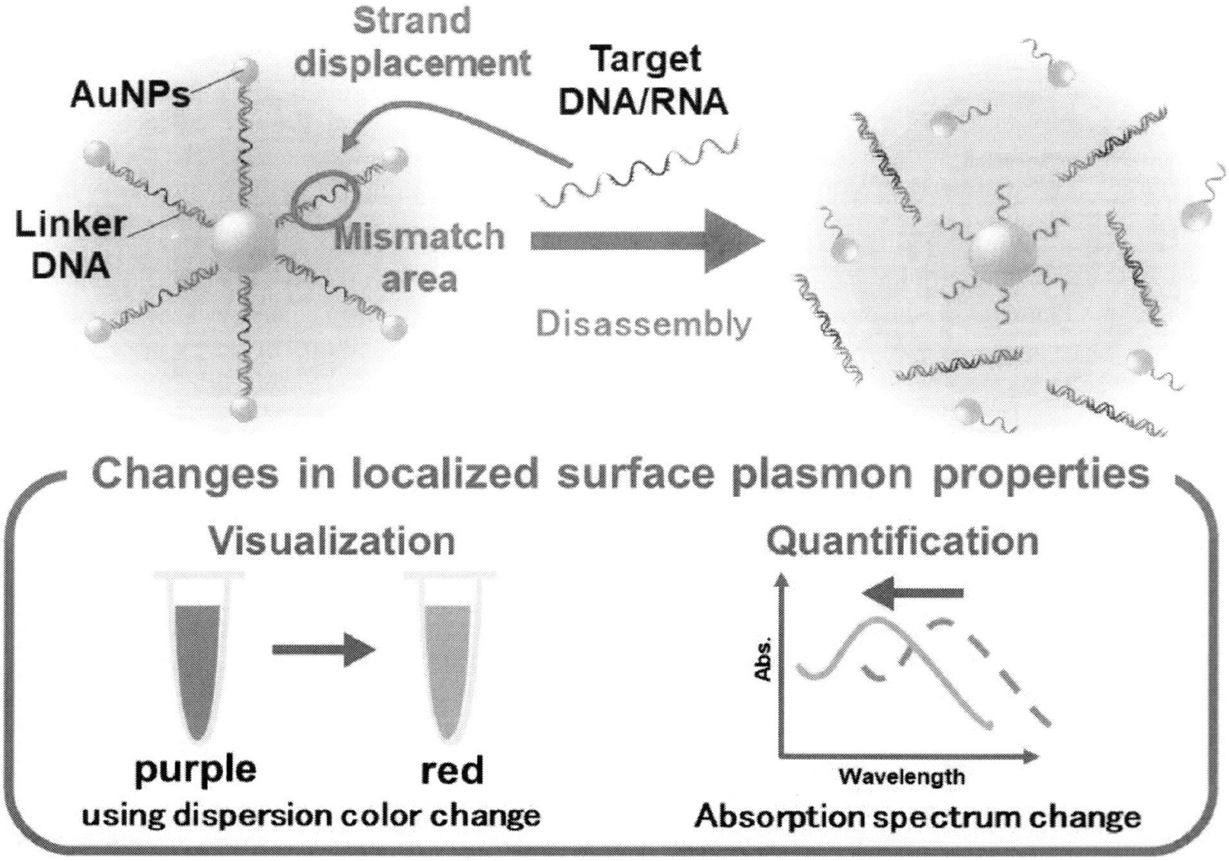
Strand
displacement
Target
DNA/RNA
AuNPs
Linker
DNA
Mismatch
area
Disassembly
Changes in localized surface plasmon properties
Visualization
Quantification
Abs.
Wavelength
purple
red
using dispersion color change
Absorption spectrum change

352c Detection and Visualization of HER2 Clustering Using an Aggregation Induced Emission Based Fluorescent Probe Molecule

Yasaman Moradi, Jerry S.H. Lee, Andrea M. Armani

The spatio-temporal organization of proteins within the cell membrane can influence basic cell behaviors including signaling and proliferation which play a role in disease initiation and progression. Understanding these control mechanisms can motivate new therapeutic designs. Several methods exist for studying the organization of membrane proteins including electron microscopy, super-resolution optical microscopy methods, and proximity-based assays. However, these methods are either not compatible with living cells or require access to specialized instrumentation. Furthermore, the highest-resolution approaches that can detect the spatial organization of proteins rely on single-molecule imaging methods and cannot be utilized in a high-throughput format.

To overcome many of the limitations of the current methods, we have developed a fluorescent microscopy tool based on a family of fluorophores that exhibit Aggregation Induced Emission (AIE) behavior for studying the spatial organization of membrane-bound proteins. Fluorophores with the AIE property can go through a fluorescence turn-on process when their intramolecular motions are restricted. For example, when AIE molecules form solid aggregates, the emission increases, and this increase in signal can be easily detected using conventional fluorescence microscopy. When proteins cluster or are in close proximity, they behave as a biological mimic for aggregation. Therefore, a tool based on an AIE fluorophore can overcome many limitations of previous methods.

In this work, we demonstrate the development and utilization of an AIE-based tool to study the response of the breast cancer biomarker HER2 to a conventional targeted cancer therapeutic (Trastuzumab). Trastuzumab (Herceptin), an FDA-approved monoclonal antibody therapeutic, is hypothesized to change the biophysical distribution of HER2 in the membrane, disrupting clusters or decreasing densities. The dependence of AIE luminogen's (AIEgen's) fluorescent response on its proximity to other AIEgens can provide insight into this mechanism.

Our AIE-based tool is developed by designing, synthesizing, and characterizing a molecule based on Tetraphenylethylene (TPE) (an AIE fluorophore) and binding it to an antibody specific to the extracellular domain of HER2 through the non-selective NHS Ester-Amine bioconjugation reaction. The target specificity of the AIE-based tool is confirmed through a competing colocalization assay using a HER2+ cell line. Then, the AIE-based tool was utilized for visualizing HER2 clusters in Her2+ cells with the presence of HER2 clusters in comparison to HER2- cells with a more uniform distribution of HER2. Since the molecular motions of the TPE domain of the tool become restricted in the cellular regions where receptors are in proximity of each other, a significantly higher fluorescent signal was observed in the HER2+ cells compared to HER2- cells.

Furthermore, the developed AIE-based tool was utilized in an assay for detecting the effect of Trastuzumab on HER2 clusters. Fluorescein-HER2 antibody, a control that is not sensitive to proximity, and the AIE-based tool are applied to HER2+ cells post-treatment with a range of Trastuzumab concentrations for different time points. The results determine that, unlike the control Fluorescein-HER2 antibody, the AIE-based tool demonstrates a significant reduction in the average fluorescent intensity of the cells treated with higher Trastuzumab concentrations and for longer. This reduction can directly correlate with the disruption of HER2 clusters or the decrease of HER2 densities.

Current results confirm the applicability of the AIE-based tool for visualization of spatial organization of HER2 and set the stage for generalizability of the tool for studying other biological systems with applications in developing improved therapeutics in the future.

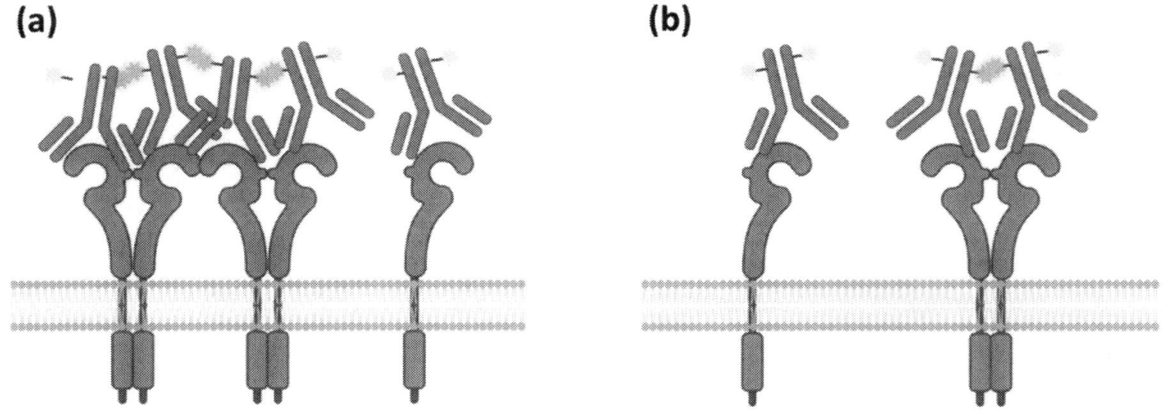

Schematic of AIE-based fluorophore for localized detection and visualization of HER2 clustering on **(a)** HER2 overexpressing and **(b)** HER- cell surfaces

352d DNA-Mediated, on-Membrane Sequential Assemblyof Conjugated Polymer Nanoparticles for Sensitive Detectionof Cell Surface Markers

Yuki Maeda, Noriko Nakamura, Seiichi Ohta

Introduction

Cell surface markers are proteins that are expressed on cell membrane. They have been reported to be associated with diseases such as cancer. Also, they can be labeled with fluorescent probes. Therefore, they have been targets for diagnosis and therapy. Flow cytometry is a gold standard to detect these markers. Although more than several thousand markers need to be expressed on a cell to detect, some crucial markers do not meet this detection limit. Therefore, improvement of detection sensitivity is necessary. To amplify fluorescent signals, in previous studies, antibody was modified with multiple fluorescent dyes. However, excessive modification made antibodies lose their binding ability. On the other hand, use of large fluorescent particles (Ca. >60 nm) with high bright fluorescence tends to inhibit the access to neighboring makers due to steric hindrance. Therefore, alternative strategy for fluorescent amplification for antibodies is needed. Conjugated polymer nanoparticles (Pdots) are one of the new fluorescent probes [1]. Pdots have been investigated in the biomedical field because of their highly bright fluorescence, excellent biocompatibility, photostability, and size-tunability. Recently, DNA-mediated assembly of nanoparticles has attracted attention. Modification of single-stranded DNA on nanoparticle surface and subsequent hybridization with linker DNA that has complementary sequences enables flexible arrangement of nanoparticles in superstructure. Since the properties of photo-functional nanoparticles such as gold nanoparticles and quantum dots change with steric configuration, such as the inter-particle distance, their assembly can be expected to expand their functions [2]. It is expected that employment of this technology for Pdots could enable highly bright labeling. However, to the best of our knowledge, there is no report on not only the assembly of Pdots using DNA, but also fluorescent nanoparticle assembly on cell membrane. In this study, we aim to achieve highly sensitive surface marker detection via sequential on-membrane assembly of Pdots mediated by DNA.

Experimental

1. Synthesis of Pdots

Pdots was synthesized by nanoprecipitation method previously reported by our group [3]. Poly [2-methoxy-5-(2-ethylhexyloxy)-1,4-phenylenevinylene] (MEH-PPV) and poly (styrene-co-maleic anhydride) (PSMA) were used as fluorescent conjugated polymer and dispersion stabilizer, respectably. The mixture of MEH-PPV (0.04 mg/mL) and PSMA (0.01 mg/mL) dissolved in tetrahydrofuran (THF) was added to 100 mL of pure water under sonication and vigorous stirring to form a homogenous Pdots solution. After purification by ultracentrifugation, the particle size distribution and morphology were evaluated by DLS and TEM observation.

2. DNA modified Pdots (DNA-Pdots)

The Pdot surface were modified with DNA by carbodiimide reaction. Amine-functionalized single stranded DNAs, 1-ethyl-3-(3-dimethylaminopropyl) carbodiimide hydrochloride (EDC) and N-hydroxy succinyl imide (NHS) were added to the Pdots dispersion for the reaction. After centrifugation to remove free DNA, surface evaluation was conducted by FT-IR and gel electrophoresis.

3. Study of Pdot assembly on cells

CD19 expressed on Nalm-6, a cell line derived from human B-cell acute lymphoblastic leukemia cells, was used as a target for demonstrating Pdot assembly on cells membrane. First, Nalm-6 cells were treated with 4 % paraformaldehyde solution. Then, DNA conjugated anti-CD19 antibody were added to cell suspension in PBS containing 1 % BSA. After the purification, linker-DNA and DNA-Pdots that have complemental sequences were added one by one, followed by the hybridization at room temperature to label CD19 with

Pdots. Afterwards, linker-DNA and DNA-Pdots were further added sequentially to achieve on-membrane Pdot assembly.

Results and Discussion

1. Synthesis of Pdots

TEM image revealed that the obtained nanoparticles have an average diameter of 28 nm. Absorption and fluorescent spectrum of the obtained MEH-PPV Pdots were also evaluated by UV-vis spectroscopy and fluorometer. Pdots exhibit broad absorption band ranging from 400 nm to 550 nm, which is convenient for fluorescent microscopy and laser excitation. The maximum fluorescent emission peak was 590 nm, which is longer than autofluorescence of cells. These results illustrated the proper optical property of Pdots for surface marker detection.

2. DNA modified Pdots (DNA-Pdots)

FT-IR spectra of free DNA and Pdots before and after DNA modification were measured. Spectrum of DNA-Pdots had peaks at 1090 cm^{-1} and 1200 cm^{-1} derived from the phosphate group of DNA, and the peak at 3400 cm^{-1} derived from the amide bond. Gel electrophoresis was also conducted in agarose gel for Pdots before and after DNA conjugation. After the conjugation of DNA, position of the band shifted to downstream region, which is caused by the increased negative surface charge. These results suggested successful DNA modification on the surface of Pdots.

3. Study of Pdot assembly on cells

To investigate if it is possible to assemble Pdots on actual cell surface markers, CD 19 expressed on Nalm-6 cells was used as a target. Confocal laser microscope was used for obtaining fluorescence image of Pdot assembly on cells at each step. Fluorescent intensity was weak at the initial step, whereas it was enhanced with the number of assemblies. Therefore, Pdots assembly on Nalm-6 cells were suggested. Fluorescent evaluation was further conducted by flow cytometry. The fluorescent signal in flow cytometry was increased with increasing number of assembly steps, demonstrating the successfully fluorescent amplification by sequential Pdot assembly on cell surface markers. By optimizing the reaction conditions for the assembly, up to 22-fold fluorescent intensity increase was achieved.

To further investigate the advantage of sequential assembly, we compared the results with one-step labeling with larger Pdots. Because the average number of Pdots (28 nm) after 2 step assembly on a marker was estimated to be 22, 95 nm Pdots, whose volume is almost equivalent to that of total assembled 28 nm Pdots, was chosen for the comparison. As a result, fluorescent intensity obtained by one-step labeling using 95 nm Pdots was much lower than that by sequential 28 nm Pdot assembly, which would be due to the steric hindrance between particles. These results further highlighted the usefulness of the proposed sequential assembly strategy of Pdots mediated by DNA.

Summary

Pdots and antibodies were modified with single stranded DNAs. After sequential assembly of Pdots on cell membranes via complementary DNA linkers, the fluorescent signal in flow cytometry was successfully amplified, up to 22-fold compared with the first step, giving a much stronger signal than conventional fluorescent dye-modified antibody. The obtained results are expected to contribute for early disease diagnostics.

References

[1] C. Wu, J. McNeill, *ACS Nano*, 2, 11,2415－2423, (2008).

[2] S. Ohta, D. Glancy, W.C.W. Chan, *Science*, 351, 841－845, (2016).

[3] N. Nakamura, N. Tanaka, S. Ohta, *RSC Adv.*, 12, 11606－11611, (2022).

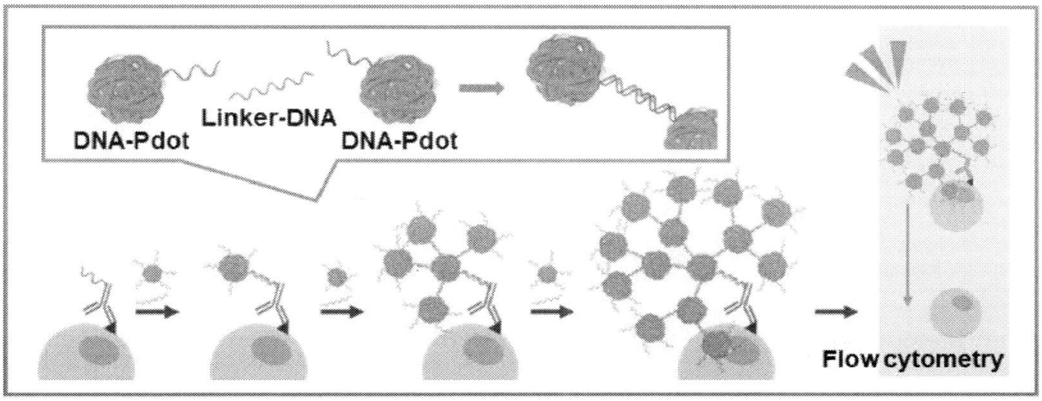
DNA-Pdot
Linker-DNA
DNA-Pdot
Flow cytometry

352e Improve Sensitivity and Selectivity of Neurotransmitter Sensing through UV Plasmonics Enhanced Native Fluorescence.

Mohammad Mohammadi, Yunshan Wang

Neurotransmitters (NTs) are responsible for brain neuronal communications and their dysregulation can lead to a number of disorders such as Parkinson's disease, Alzheimer's disease, depression and addiction. The ability to simultaneously assay multiple NTs and their metabolites at their physiological concentration is essential to understand their functional roles, the relations between these chemicals and develop interventions that may normalize their levels in pathological conditions. Conventional technologies for NT detection include electrochemical methods. Electrochemical methods cannot probe multiple NTs simultaneously in one region and cannot differentiate structurally similar molecules such as dopamine (DA) and norepinephrine (NE).

This work presents a neurotransmitter sensor based on ultraviolet plasmonic enhanced native fluorescence that promises to detect an array of NTs with high sensitivity and selectivity. Monoamine NTs have intrinsic fluorescence in the ultraviolet range of the spectrum. However, the intrinsic fluorescence intensity is low compared to fluorescence dyes. UV plasmonics has been used to enhance the native fluorescence of biomolecules. We report that UV plasmonics not only improve the detection limits by enhancing the native fluorescence; it also improves the selectivity by providing a new mechanism for molecule differentiation. Three neurotransmitters have been tested: dopamine, norepinephrine, and 3,4-dihydroxyphenylacetic acid (DOPAC). The net enhancement on an aluminum hole array with 300 nm hole spacing substrate was measured to be 50 times for all three molecules. The photobleaching rates for the 3 molecules are distinct: norepinephrine has the highest photobleaching rate, followed by dopamine and DOPAC, providing a novel differentiation mechanism. This study describes evidence that the detection and differentiation of neurotransmitters with similar structures can be achieved by UV plasmonic enhanced native fluorescence.

352f Field Flow Fractionation-Based Analysis for Evaluating Slight Aggregation of Surface-Functionalized Nanoparticles That Can Impact Their Biological Interaction

Hiroki Tsuchiya, Noriko Nakamura, Seiichi Ohta

Introduction

Surface modification of nanoparticles (NPs) has been widely studied for biomedical applications, e.g., imaging, drug delivery, therapy, and sensing. Various functional molecules such as antibodies and DNAs have been modified to NPs for mediating their biological interaction. During these surface modifications, aggregation of NPs frequently occurs, leading to loss of their functions. Therefore, special efforts have been made to prevent the aggregation of NPs, where evaluation of dispersion stability is particularly important.

Recent studies have revealed that physicochemical property of NPs, including size, shape, and surface chemistry, significantly affect their biological interactions[1]. Because aggregation of NPs dramatically changes their physicochemical properties, their interaction with biological environment, such as cellular uptake, was reported to be altered depending on the degree of aggregation.[2,3] From these studies, an emerging concern is that although NPs seem to be monodispersed, slightly aggregated particles contained in the dispersion, such as NP dimers and trimers, could significantly impact their biological fate. However, this effect has never been addressed due to the lack of an exact evaluation method. Slight aggregation possibly occurs during all the conventional surface modifications, therefore a method that can detect it needs to be developed.

One of the common evaluation methods for aggregation is Dynamic Light Scattering (DLS). DLS is a method for estimating the hydrodynamic size of NPs from fluctuations in light scattering from particles, which has the advantage of simplicity. However, because curve fitting is performed assuming spherical particles and certain particle size distributions, it is difficult to detect slight aggregation. Furthermore, since DLS measures a hydrodynamic size, the measured particle size increases due to the surface modifications, making it difficult to distinguish it from slight aggregation.

In this study, we developed a new evaluation system for analyzing slight aggregation of NPs based on Field-Flow Fractionation (FFF). Centrifugal FFF (CF3) is a method for size classification of NPs using a balance between diffusion and centrifugal force[4]. Since density is a dominant factor in centrifugal force, NPs with high density, such as metal or semiconductor NPs, are expected to be classified based on the size of their solid contents with negligible influence of the surface chemistries. Therefore, it is expected that this characteristics of CF3 would make it feasible to detect the slight aggregation of NPs. Using a sample prepared by a conventional method that could cause slight aggregation as a model, the difference on cellular uptake between slightly aggregated and monodispersed sample was investigated by CF3-based evaluation. The impact of the slight aggregation on the biological interaction of NPs will be examined via cellular uptake assay.

Experimental

Preparation of slightly aggregated NPs

Fluorophore-conjugated antibody (Ab) was modified with NHS- and SH- terminated polyethylene glycol (PEG) (MW=5000) and then modified to gold nanoparticle (AuNP) surface. During Ab modification to AuNP, SH-terminated shorter PEG (MW=2000) was also added to stabilize them. For purification of Ab-modified AuNPs, centrifugation have been conventionally used, although it could potentially cause aggregation due to their forced packing. In this study, slightly aggregated sample was prepared on purpose using relatively strong centrifugal force (15,000×g, 3 times). Synthesis of Ab-modified AuNP was confirmed by FT-IR spectroscopy. Size distribution of slightly aggregated and monodispersed Ab-modified AuNPs were evaluated by both DLS and CF3 analysis, followed by their comparison.

Morphologies of each fraction obtained from CF3 analysis were also observed by TEM.

Cellular uptake study

To examine the difference in cellular uptake between slightly aggregated and monodispersed samples, anti-EGFR Ab was modified to AuNP by the same method with above. Then, we incubated anti-EGFR Ab-modified AuNPs with A431 cells for 6 hours, followed by washing with PBS. After digesting the cells, the amount of AuNPs internalized into the cells was evaluated using ICP-OES.

Results and Discussion

Ab modification on AuNPs

In both slightly aggregated and monodispersed samples, same peaks including broad bands on 3300 cm^{-1}, which correspond to amide bonds of Ab, were observed by FT-IR spectroscopy. These results suggested successful modifications of Ab on AuNP surface.

Size evaluation via CF3 to distinguish slight aggregation

Size distribution of slightly aggregated and monodispersed Ab-modified AuNPs was evaluated by both DLS and CF3. DLS analysis showed that compared with bare AuNP, hydrodynamic sizes of Ab-modified AuNP were increased in both slightly aggregates and monodispersed samples. However, the difference between them was not obvious enough to identify the slight aggregation. On the other hand, in CF3 analysis, the main elution peak of monodispersed Ab-modified AuNP sample was same with that of bare AuNP. However, elution peak of slightly aggregated Ab-modified AuNP showed distinct multiple peaks, corresponding to AuNP dimers and trimers as evidenced by TEM observation.

The observed difference between DLS and CF3 analysis would be attributed to the different evaluation mechanism. Because DLS detects hydrodynamic size of NP, measured particle sizes increase due to not only slight aggregation of NPs but also surface modification to NPs, making it difficult to distinguish them. On the other hand, because CF3 classifies NPs basically based on the mass of NPs, even if the surface was modified with Ab and PEG, the total mass changes were small enough to be considered as negligible because of the high density of AuNPs themselves. Therefore, CF3 could classify monodispersed and slightly aggregated Ab-modified AuNPs based on the number of NPs bound together. These results demonstrated that CF3 could detect slight aggregation of Ab-modified AuNPs, which was not detectable by conventional DLS analysis.

Effect of slight aggregation on cellular uptake

So far, we proposed a new evaluation system for slight NP aggregation using CF3. Herein, we show how problematic the presence of slight NP aggregation could before biomedical application. To evaluate the difference between slightly aggregated and monodispersed Ab-modified AuNPs in their biological interaction, the cellular uptake amount by A431 cells was compared. The cellular uptake level was 1.8-times higher for slightly aggregated sample than monodispersed sample ($p < 0.001$). This trend was consistent with previous reports on receptor-mediated endocytosis in this size range, around 50 nm; larger particles have more Abs on surface and thus bind more frequently to the receptor than smaller one[5]. Since a mere presence of slight aggregation could change the amount of cellular uptake as shown above, attention and removal of these slight aggregation would be required to obtain more accurate and reproducible results. These results demonstrated the impact of slight aggregation of NPs on their biological interaction and the importance of its evaluation via CF3.

Conclusions

CF3 was able to detect slight aggregation of Ab-modified AuNPs prepared via a generally used method, which was difficult to be detected by DLS. It was demonstrated that the slight aggregation of Ab-modified AuNP significantly affect their cellular uptake behavior. For more accurate NP-based biomedical applications, evaluation via CF3 would be important.

Acknowledgement

This work was joint research with SHIMADZU CORPORATION and supported by JST PRESTO.

References

1. Chou, L. Y. T., Ming, K. & Chan, W. C. W. Strategies for the intracellular delivery of nanoparticles. *Chem Soc Rev* **40**, 233–245 (2011).
2. Albanese, A. & Chan, W. C. W. Effect of gold nanoparticle aggregation on cell uptake and toxicity. *ACS Nano* **5**, 5478–5489 (2011).
3. Liu, X. *et al.* Enhanced retention and cellular uptake of nanoparticles in tumors by controlling their aggregation behavior. *ACS Nano* **7**, 6244–6257 (2013).
4. Williams, P. S. & Giddings, J. C. Power Programmed Field-Flow Fractionation: A New Program Form for Improved Uniformity of Fractionating Power. *Anal Chem* **59**, 2038–2044 (1987).
5. Chithrani, B. D., Ghazani, A. A. & Chan, W. C. W. Determining the size and shape dependence of gold nanoparticle uptake into mammalian cells. *Nano Lett* **6**, 662–668 (2006).

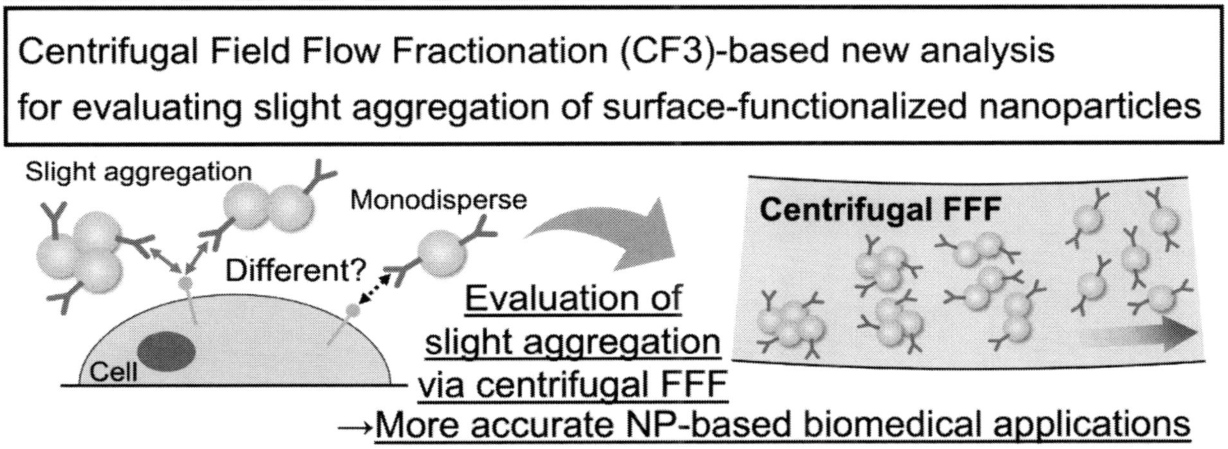

352g An Injectable Hyaluronic Acid-Based Nano-Contrast Agent for CT Detection of Inflammation and Osteoarthritis

Ioanna Chatzaki, Alice Truong, Mary Ruskowski, Manos Gkikas

Osteoarthritis (OA) is a common and painful chronic joint disease that affects large joints such as the knees, spine, hips, and shoulders. It is the most prevalent chronic joint disease in the world, with half of the world's population aged 65 or older suffering from some form. ***OA relates to the decomposition of hyaluronic acid (HA) in the joints***, a high molecular weight non-sulfated poly(glycosaminoglycan) and the main component of synovial fluid, a viscous (physical) hydrogel that lubricates the joint surfaces and reduces friction between articular cartilages with movement. Current diagnosis involves imaging with MRI or X-rays. An FDA-approved treatment includes ***HA injections in the knee*** of OA patients (~2 mL of HA once a week; for 3-5 weeks), providing pain relief for up to 6 months, and is recommended when patients cannot control pain with analgesic drugs or tolerate those drugs. We develop a *minimally-invasive,* ***diagnostic HA-based nano-contrast agent*** that can be administered intra-articularly in the knee of OA animals and retained, similar to FDA-approved HA, and can provide a ***more sensitive detection of OA / joint injuries by X-ray CT*** (computed tomography). The injectable nano-scaffold can hinder bone friction, similarly to viscosupplemented HA, and simultaneously ***quantify joint degradation and inflammation*** through X-rays. The utilized method takes advantage of reactive oxygen species (ROS) produced in inflamed joints. The presence of those in conjunction with our nano-scaffold leads to ***X-ray CT signal enhancement***. Among our achievements is the *low amount of metal (<1 mg)* that is utilized, and the use of *photon-counting spectral CT* (PCD-CT). Synthesis and characterization of the material will be shown, *in vitro* studies with ROS and signal enhancement, as well as results from our first *in vivo* studies.

229a Mucus Penetration of Surface-Engineered Nanoparticles in Various pH Microenvironments.

Yiyang Guo, Bingbing Sun

The inhaled medications delivered by nano- or micro-sized carriers are widely used for the prevention and treatment of lung diseases such as cystic fibrosis (CF), chronic obstructive pulmonary disease (COPD), asthma, and COVID-19. During the transport of the drug carriers in the airway, the mucus lines the surface of the airway and hinders the effective delivery of active ingredients. The mucus layer is composed of 98% water, 1.6% mucin, and 0.4% of the remainder, *e.g.*, cellular debris, salts, DNA, lipids, globular proteins, *etc.* Mucin form a three-dimensional network to restrain the movement of particles *via* the size exclusion and interaction filtration effect. It has been demonstrated that proper control of physicochemical properties of carriers could improve the transport and fate of medications.

The neutral charge and hydrophilic surface have been identified as critical requirements for the effective trans-mucus drug carriers. Poly (ethylene glycol) (PEG) has been used as a stealth polymer that repels protein adsorption to assist the drug carriers to escape the binding of mucin. Their molecular weight, chain length, and content have been shown to affect mucus penetration performance. The zwitterionic design mimics the surface characteristics of the virus surface, and further exhibits the equal densities of positive and negative charges without the presence of hydrophobic patches. The surface modification by zwitterionic polymer and density adjustment enables enhanced mucus penetration of particles. Furthermore, the particle surface modification by block copolymer (F-127), peptides and protein were potentially alternative strategies to improve penetration ability of particles. Although PEGylation is often considered as a "gold standard" to assess mucus penetration, and various surface modification strategies have been proposed. the mechanistic understandings of particle surface property-mediated mucus penetration are currently missing.

The mucus microenvironment changes according to the physiological and pathological states. In asthma exacerbation, healthy, and airway infection or inflammation states, the pH values of mucus is altered. However, the airway mucus penetration model in current study is mainly derived from commercial porcine gastric mucin. Regardless of the mucin suspension or the artificial mucus, the pH value of the system is acidic (pH=3-5), which is different from the natural airway health neutral mucus environment. Although sputum taken from patients such as cystic fibrosis (CF) as a surrogate is similar to the mucus environment experienced by drug carriers after airway administration, factors such as the causes (endogenous or exogenous exposure) and states of diseases (mild or severe) were not considered. Acid respiration occurs in exacerbation of CF (mucus pH=5.3) and asthma (pH=5.2), but the pH of mucus is increased in mild asthma (pH=7.6) and stable CF (pH=5.9). Moreover, the mucus is an alkaline environment (pH=7.8-8.5) in inflammatory diseases. Therefore, mucus pH microenvironment should be taken into consideration during inhalation delivery.

In this study, we prepared a comprehensive library of engineered silica nanoparticles (SNPs) with controlled surface ligand type and density. According to the size exclusion mechanism of mucus, nanoparticles with a size range of 200-500 nm in diameter are easier to penetrate the mucus barrier. Therefore, the 300 nm SNPs were selected to study their mucus penetration ability. The pristine SNPs were synthesized using the Stöber method, and SNPs were doped with fluorescein isothiocyanate (FITC) for particle tracking. Then, SNPs were further modified with silane coupling agents to prepare a library of engineered SNPs with controlled functional groups and ligand densities. Then, particles were dispersed in the airway mucus with mimetic physiological and pathological pH values to evaluate particles movement behavior by using multiple-particle tracking (MPT). PEGylated and amine SNPs showed pH-independent mucus penetration and entrapment, the carboxyl-SNPs exhibited movement enhancement only in weakly alkaline environments. The mechanisms were further systematically analyzed from three aspects, *i.e.*, mucus properties, SNPs surface properties, and mucus-SNPs interactions, respectively.

A recent study has revealed that mucin undergoes a sol-gel transition in acidic-neutral environments, and the cross-linking properties of mucus are stronger in acidic than that in neutral condition. In our study, it was demonstrated that the pH-dependent changes in mucus macro-rheology and micro-structure were the main reasons for the observed variations of particle motion behavior. From the perspective of mucin micro-structure, the changes in the macro-crosslink properties of mucus at different pH was due to the changes in the microstructure of mucin. During the acidification of mucus, carboxylates on mucin were protonated. As a result, the electrostatic interaction between carboxylate and amine groups were compromised, and hydrophobic domains hidden in salt bridge folded structure were exposed, promoting the cross-linking of mucus gel. In addition to the influence of the hydrophobic domain, electrostatic interactions due to the exposure of mucin glycosyl side chains also play an essential role. In generally, mucin is negatively charge due to the sialic and sulfate components in the glycosyl side chain. In the present study, this pH-dependent conformation and charge transition were observed. The circular dichroism (CD) analysis revealed that mucin was dominated with the random coil conformation, and the pH-dependent intensity and width changes at 206 nm was observed, suggesting the conformation undergoes a transition from isotropic random coil to anisotropic extended random coil with the decrease of pH values. Additionally, the mucin fiber in acidic environment exhibited increased zeta potential, thus enhancing electrostatic interactions between mucin fibers, increasing the viscosity of the mucus and inhibiting the movement of the particles.

Furthermore, isothermal titration calorimetry (ITC) reveals that the independent of the functional groups and mucus pH, the interaction of mucin with SNPs were exothermic and entropy loss process, indicating an enthalpy-driven interaction and the formation of non-covalent bonds was reason why particles were trapped in mucus, while the dense brush structure formed by PEG modification shielded the interaction between particles and mucin to achieve mucus penetration. Furthermore, to study the effect of mucus pH, the amine-modified SNPs were selected to titrate with mucin solutions. The association constant (K_a) of SNPs in acidic (pH 5.2) environment was an order of magnitude higher than those in neutral (pH 7.0) and slightly alkaline (pH 7.7) ones. The stronger SNP-mucin interaction was correlated with the fact that particles being trapped in the in acidic mucus. Therefore, this study demonstrated that the fate of drug carrier particles in airway mucus was pH microenvironment, particle surface ligand type, and density-dependent, and provides an engineered approach to control the mucus penetrating capabilities of drug delivery carrier for the personalized treatment and prevention of pulmonary diseases.

Abstract figure legend:

Airway mucus penetration and mechanism analysis. (a) Scheme showing the preparation of surface engineered silica nanoparticles synthesis routes. (b) TEM image of pristine and surface-modified silica nanoparticles. The scale bar is 300 nm. (c) Zeta potentials of silica nanoparticles in water. (d) Scheme showing the multiple-particle tracking (MPT). (e) The viscosity of artificial mucus at different pH as a function of shear rate in the range of $1\text{-}10^3\ s^{-1}$. (f) Ensemble mean squared displacement of surface engineered silica nanoparticles in acidic, neutral, and alkaline artificial mucus and (g) native rat ileal mucus. (h) Distribution of surface engineered silica nanoparticles in the airway of mice. (i) Schematic illustration of SNP-mucin interaction revealed by using isothermal titration calorimetry (ITC).

Modulating Nonspecific Uptake of Engineered Extracellular Vesicles

Authors: Beth DiBiase[1]*, Roxana Mitrut[1], Taylor Gunnels[1], Neha Kamat[1], Joshua Leonard[1]
[1]Northwestern University, Evanston, IL

Engineered extracellular vesicle (EV) therapies are emerging treatment modalities for many diseases. An open challenge for clinical use is the susceptibility of EVs to nonspecific uptake by the mononuclear phagocytic system (MPS), leading to fast clearance times[1]. Extending circulation time and decreasing nonspecific uptake would provide therapeutics with more opportunities to reach their targets while maintaining the biological advantages of EVs as delivery vehicles. Several strategies for reducing nonspecific uptake of EVs have proven effective, such as decorating the surface of EVs with polyethylene glycol (PEG) as a shielding moiety[2]. However, this can result in the formation of anti-PEG antibodies and other negative immune responses, emphasizing the need for a therapeutic delivery vehicle which can evade the MPS while maintaining safety and efficacy. We hypothesized that engineering EVs with biopolymers, a potentially safer and more efficacious shielding molecule, would reduce their nonspecific uptake and increase circulation time, while providing a tunable handle for modulating these properties to fit the application. To evaluate our hypothesis in vitro, we isolated EVs from mScarlet fluorescent FreeStyle 293-F cells (suspension cell line) engineered to display varying densities of a shielding biopolymer on their plasma membranes using ultracentrifugation. Measurement of biopolymer loading on EV surfaces was determined via fluorescent proxy, and biopolymer EV morphology is currently being evaluated via cryo-EM. After an in vitro macrophage uptake assay, changes in nonspecific uptake of fluorescent EVs were measured by flow cytometry. RAW 264.7 mouse macrophage-like cells were utilized in this assay, as they provide an indication of the potential effects on nonspecific uptake by the MPS in vivo. While ultracentrifuged mScarlet fluorescent EVs are suitable for evaluation in the in vitro context, nanoluciferase (Nluc) luminescent EVs isolated via tangential flow filtration are more practical for scale up and visualization in the in vivo context. Future studies will utilize Nluc luminescence of EVs modified with biopolymer shielding molecules from blood draws at multiple time points to investigate circulation time in vivo. Preliminary uptake study results have indicated a reduction in macrophage uptake compared to control conditions when EVs are loaded with a mucin biopolymer (Figure 1). We theorize that this decrease in nonspecific uptake can be modulated by tuning the amount of biopolymer displayed on the EVs. Further investigation into balancing EV shielding with targeting is ongoing. Studies investigating the changes in membrane properties which affect uptake will elucidate the underlying mechanisms, and preclinical studies investigating circulation time will advance therapeutic treatments. We anticipate that this study will expand the field of engineered EVs therapeutics, accelerating their development to combat diseases.

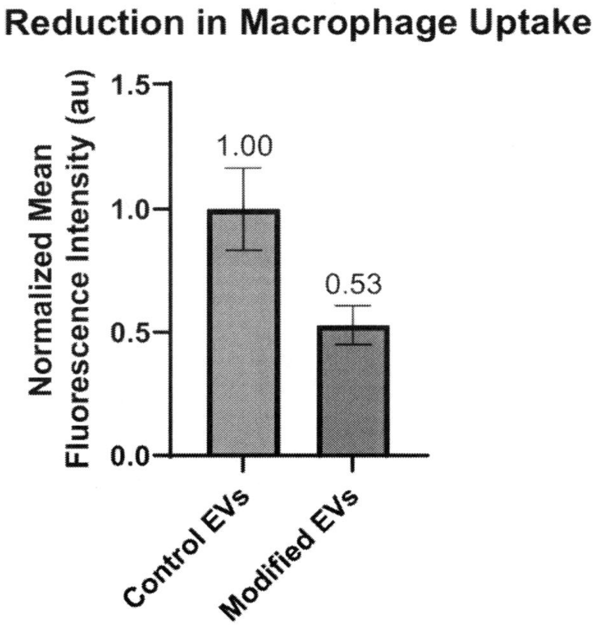

Figure 1. Biopolymer shielding moiety loaded on EVs leads to reduced uptake of EVs by macrophages, as represented by a decrease in MFI evaluated via flow cytometry.

Acknowledgements: This research was supported by the National Science Foundation (DGE-2234667). Any opinion, findings, and conclusions or recommendations expressed here are those of the authors and do not necessarily reflect the views of the NSF.

References:

1 Belhadj, Z. *et al.* A combined "eat me/don't eat me" strategy based on extracellular vesicles for anticancer nanomedicine. *Journal of Extracellular Vesicles* **9**, 1806444 (2020). https://doi.org:10.1080/20013078.2020.1806444

2 Kooijmans, S. A. A. *et al.* PEGylated and targeted extracellular vesicles display enhanced cell specificity and circulation time. *Journal of Controlled Release* **224**, 77-85 (2016). https://doi.org:10.1016/j.jconrel.2016.01.009

229c Bubble-Based Microrobots for Epithelial Pinning and Sustained Drug Delivery

Jin Gyun Lee, Ritu Raj, Cooper P. Thome, Nicole Day, Payton Martinez, Nick Bottenus, Ankur Gupta, C. Wyatt Shields IV

Remotely powered active colloids have been recognized as next-generation medical microrobots for drug and gene delivery. However, most microrobots swim with linear trajectories and lack the capacity to robustly adhere to soft tissues. This limits their ability to navigate complex biological environments and sustainably release drugs at target sites. In this work, bubble-based microrobots with complex geometries are shown to efficiently swim with non-linear trajectories in a mouse bladder, robustly pin to the epithelium, and slowly release therapeutic drugs for sustained effects. The asymmetric fins on the exterior bodies of the microrobots induce a rapid rotational component to their swimming motions of up to ~150 body lengths per second. Due to their fast speeds and sharp fins, the microrobots can mechanically pin themselves to the bladder epithelium and endure shear stresses commensurate with urination. Dexamethasone, a small molecule drug used for inflammatory bladder diseases, is encapsulated within the polymeric bodies of the microrobots. The sustained release of dexamethasone is shown to temper inflammation in a manner that surpasses the performance of free drug controls by comparing the molecular expression of key inflammatory markers on murine macrophages. This system provides a potential strategy to use microrobots to efficiently navigate large volumes, pin at soft tissue boundaries, and release drugs over several days for a range of diseases.

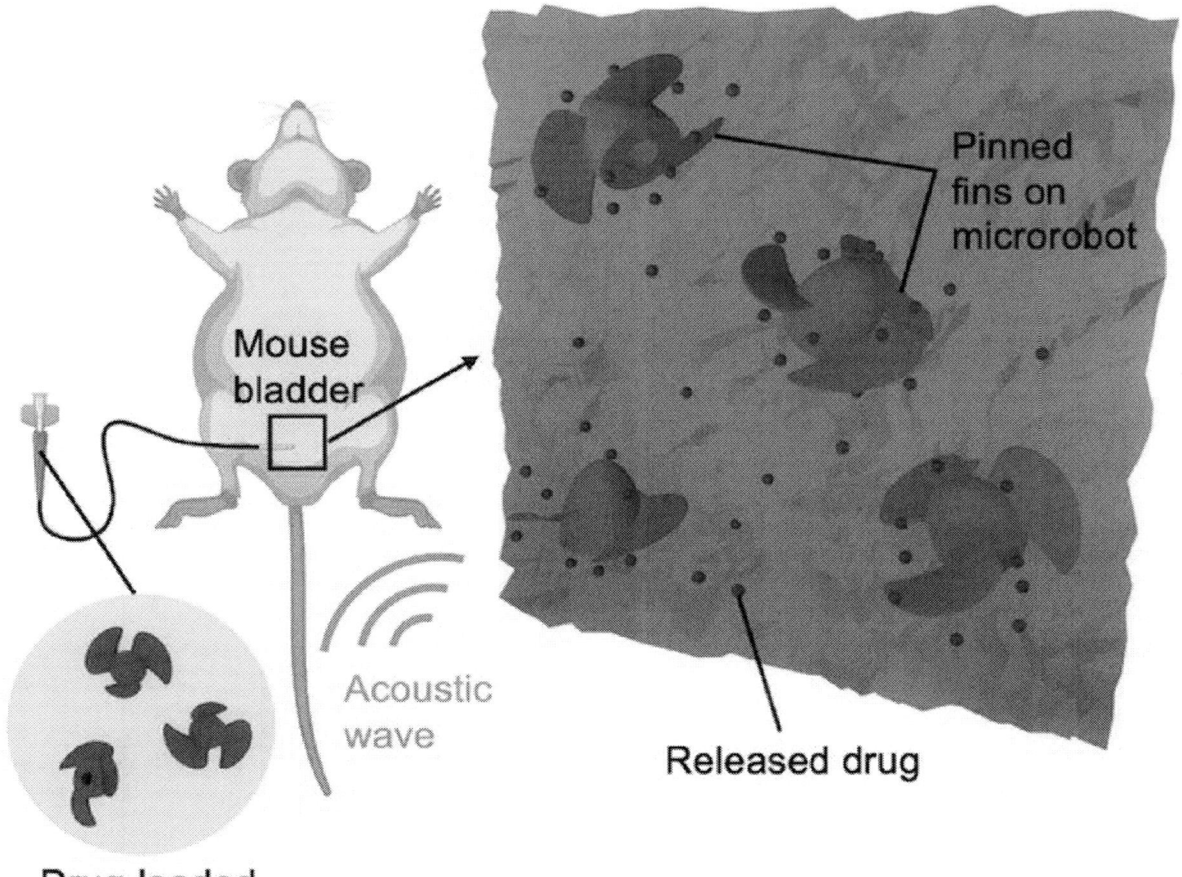

229d On-Demand Drug Release of Polymeric Core-Shell Nanoparticles Developed through Aqueous Surface-Initiated Sara-ATRP

Nahieli Preciado, Colin Stark, Jingyi Liu, Todd Hoare

Background

Chemotherapeutic delivery vehicles that can be externally addressed to release drugs 'on-demand' offer significant potential to dynamically adjust treatment according to the patient's response (i.e. personalized medicine) without requiring repeated invasive and often painful drug administrations. Even though new drug delivery methods that lower secondary risks have been developed via nanomedicine, they have not significantly improved patients' survival rates [1]. New therapeutic paradigms are required to achieve the desired improvements in patient outcomes.

One key group of materials with potential in this regard is nanoscale stimuli-responsive systems that react to either exogenous stimuli such as temperature variations, magnetic fields, ultrasound, light, or electric fields [2] or endogenous changes in local pH and enzyme concentration associated with a disease [3]**.** Magnetic triggering is especially attractive as an external trigger since magnetic fields are highly penetrative (>1 m in all tissues [4], as opposed to light and, in specific scenarios, ultrasound, that have penetration depths as short as the millimetre scale) without adversely affecting biological tissues at biologically-relevant field strengths/frequencies. The most common implementation of magnetic triggering applies superparamagnetic iron oxide nanoparticles (SPIONs), whose single-crystal nanostructure enables the transformation of energy from an alternating magnetic field (AMF) into heat through Neel and Brownian relaxation mechanisms. SPIONs can be used alone for cancer therapy by leveraging this heat generation capacity to induce localized hyperthermia to kill cancer cells, with thermotherapy noted to have fewer side effects than chemo or radiotherapy [5]. However, increasing recent interest has focused on creating hybrid nanosystems that merge inorganic nanoparticles with drug delivery systems to **synergistically** induce apoptosis in cancer cells, with the drug acting together with the higher local temperature to create a more efficient cancer therapy [6], [7].

Introducing the potential for *pulsatile*, rather than simple diffusion-controlled, release would add another therapeutic option to trigger release of drug "on-demand" as per a patient's need, addressing the lack of control over release kinetics currently achievable once a drug delivery vehicle is administered into the body. Thermoresponsive hydrogels have been previously used for this purpose; however, their high porosity allows loaded drug to be released rapidly coupled with relatively high "off" state drug release by diffusion through even the collapsed gel phase. As an alternative, the use of the glass transition (Tg) properties of solid polymer-based particles would significantly reduce "off" state release while still allowing for temperature-induced "on" state release enhancements if the Tg is tuned to be just above physiological temperature. While current research has successfully led to the development of a range of injectable smart materials on many length scales and with many geometries, the capacity to safely and non-invasively control the smart response(s) locally and repeatedly without biological side-effects over the long term, remains challenging. By combining the glass transition with core-shell nanoparticles that can enable stabilization/targeting of the nanotherapeutic, we see the potential to develop an injectable drug delivery vehicle with the capacity for repeated pulsing while avoiding substantial drug delivery in the 'off' state that currently hinders the translation of on-demand drug delivery devices.

Objective

We aim to develop a nanoscale injectable on-demand drug delivery system that can be activated by heat and/or magnetic fields to selectively deliver drug only when triggered. Such a system will decrease the secondary effects of systemic cancer therapy drugs, increasing drug concentration at the target site, and reduce total drug dose compared to the systemic delivery approach. The gel shell around the glass transition-switchable nanoparticle promotes colloidal stability and reduces biological responses. The heat-induced trigger will allow for on-demand release of only the needed amount of drug only when heat or an alternating magnetic field is applied, the latter enabled by the incorporation of SPIONs in the shell of the particles.

Methodology

The core nanoparticles were fabricated by free radical polymerization of methyl methacrylate (MMA), butyl methacrylate (BMA), and 2-(2bromoisobutyryloxy)ethyl methacrylate BIEM) using an oil-in-water miniemulsion technique, using rhodamine B as the model drug loaded by directly incorporating it during the miniemulsion process. The particle size was determined via dynamic light scattering. Following, the bromine groups in BIEM were used as an initiator to grow the OEGMA shell via aqueous surface-initiated SARA-ATRP. SPION groups were incorporated in the shell. Zwitterionic polymers functionalized with either ketone or hydrazide polymers (enabling the formation of a hydrazone bond upon contact) were used to form an *in situ*-gelling hydrogel to immobilize the core-shell particles at a specific injection site, with the polymer concentration adjusted to obtain a desirable gelation time (1-2 min). The core-shell particles were dispersed in one or both precursor polymer solutions and then co-injected through a double barrel syringe fitted with a static mixer to form the nanocomposite hydrogels for drug release testing. Rhodamine B release was assessed by immersing the nanocomposite gels in PBS and exposing them to heat (45°C) or a 200 kHz alternating magnetic field for activation.

Results

Core nanoparticles were formulated using different ratios of monomers and macroinitiators to determine the effect the macroinitiator would have on the final glass transition temperature. Increasing the concentration of the macroinitiator from 0.25% to 1.4% only decreased the Tg by 2°C (58.5°C to 55.7°C), too high for safe triggering under physiological conditions. As such, to obtain a desirable Tg (45°C). butyl methacrylate was added to the MMA-BIEM mix, decreasing the Tg by 10°C and obtaining a final Tg of 45.6°C. The particle sizes were not significantly affected by the comonomer composition, yielding core particles of ~120 nm in diameter across a range of monomer contents.

The SARA-ATRP reaction was subsequently optimized to grow the passivating polymeric shell around the core nanoparticle. Different parameters such as the polymerization temperature and the amount of macroinitiator/BIEM were changed to determine the best conditions for shell growth. Using sub-ambient temperatures or higher concentrations of macroinitiator promoted interactions between the particles, creating a particle network rather than individual core-shell nanoparticles. In contrast, performing the reaction at room temperature and decreasing macroinitiator concentration resulted in stable core-shell nanoparticles. Shell growth was analyzed using dynamic light scattering after different polymerization times to control the shell thickness, allowing for the creation of core-shell nanoparticles of ~300 nm diameter. SPIONs were then precipitated into the shell using the xxx groups to nucleate the particle growth, with the successful incorporation of SPIONs into the shell confirmed by transmission electron microscopy.

Rhodamine B release was then tracked following encapsulation of the core-shell nanoparticles in the *in situ*-gelling zwitterionic hydrogel to immobilize the core-shell nanoparticles at a target therapeutic site. Release was tracked (via fluorescence analysis) into a PBS bath by: (1) heating the nanocomposite hydrogel from 37°C and 45°C; or (2) maintaining the bulk temperature at 37°C but applying alternating magnetic field (AMF) pulses. A 30-minute AMF pulse at 37°C achieved the same release as heating the nanocomposite to 45°C, with the 37°C/No-AMF condition facilitating significantly lower drug release. As such, AMF and temperature can both trigger a significant pulse of drug release from the core-shell nanoparticles.

Conclusion

Core-shell nanoparticles containing a core with a super-physiological glass transition temperature (Tg ~ 45°C), a stabilizing shell, and precipitated superparamagnetic iron nanoparticles in the shell can promote triggered release of drug upon direct heating or an alternating magnetic field stimulus.

Significance

We developed the first polymer-polymer core-shell nanoparticles via aqueous SI-SARA-ATRP and the first polymeric core-shell nanoparticle system used as a drug delivery vehicle, triggerable by either heating or magnetic field activation. We anticipate this system offers potential for facilitating improved pulsatile drug delivery with an improved "off" state release (taking advantage of the very slow drug diffusion through the core particle under glassy conditions), facilitating more specific on-off drug release and thus longer functionality and higher specificity of drug release at a target site.

[1] H. He, et al, "Survey of Clinical Translation of Cancer Nanomedicines-Lessons Learned from Successes and Failures," *Acc. Chem. Res.*, vol. 52, pp. 2445–2461, 2019.

[2] S. S. Said, et al, "Externally Addressable Smart Drug Delivery Vehicles: Current Technologies and Future Directions," *Chem. Mater.*, vol. 31, no. 14, pp. 4971–4989, 2019.

[3] H. Chen, et al, "Endogenous Stimuli-responsive Nanocarriers for Drug Delivery," *Chem. Lett*, vol. 45, pp. 242–249, 2016.

[4] P. J. Rosch, "Electromagnetic Fields: Biological Interactions and Mechanisms, Martin Blank, Editor, American Chemical Society, Washington, D.C., 1995, 550 pgs., $116.95," *Stress Med.*, vol. 13, no. 2, pp. 137–137, 1997.

[5] S. Sabale, et al, "Recent developments in the synthesis, properties, and biomedical applications of core/shell superparamagnetic iron oxide nanoparticles with gold," *Biomater. Sci.*, vol. 5, pp. 2212–2225, 2017.

[6] A. Hervault *et al.*, "Doxorubicin loaded dual pH-and thermo-responsive magnetic nanocarrier for combined magnetic hyperthermia and targeted controlled drug delivery applications," *Nanoscale*, vol. 8, pp. 12152–12161, 2016.

[7] W.-H. Chiang *et al.*, "Superparamagnetic Hollow Hybrid Nanogels as a Potential Guidable Vehicle System of Stimuli-Mediated MR Imaging and Multiple Cancer Therapeutics," *Langmuir*, no. 29, pp. 6434–6443, 2013.

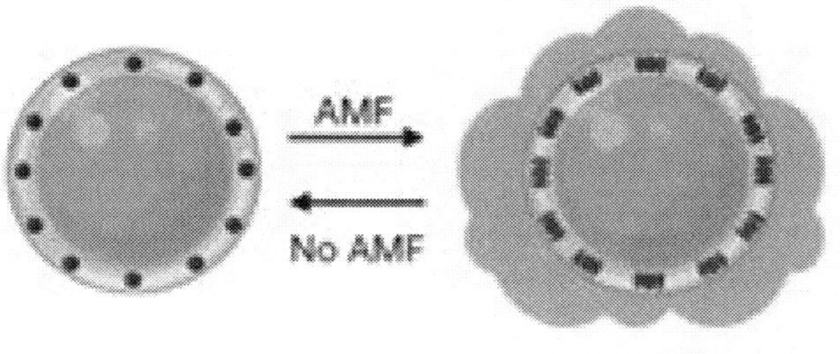

229e Supramolecular Peptide-Protein Granules for Intracellular Protein Delivery

Renjie Liu, Madisen Domayer, Alex Adolphson, Greg Hudalla

The ability to deliver active proteins across the cell membrane and into the cytosol would open up a new multi-billion dollar market by providing access to drug targets and signaling pathways that are not accessible within the extracellular environment, with faster pharmacokinetics than what are afforded by DNA and mRNA delivery approaches that require transcription and/or translation. Due to their large size, charge, and hydrophilicity, proteins do not efficiently cross the cell membrane to enter the cytosol on their own. Here we will report on the development of supramolecular peptide-protein granules for intracellular protein delivery. This approach is based on pairs of charge-complementary molecules known as, "CATCH(+) peptides" (sequence = XQKFKFKFKQX; X = Q or K), and, "CATCH(-) fusion proteins" (sequence = YQEFEFEQY; Y = Q or E), which associate in water to form β-sheet materials when combined, but remain in the soluble state when alone. Above a critical concentration of ~200 mM, CATCH(+/-) pairs form elongated fibrillar structures, while in the mM range, these fibrils can entangle to form macroscopic hydrogels. Here we will show that introducing a "crowding agent", such as Tween-80 micelles or poly(ethylene glycol), can induce the formation of 100-200 nm "CATCH(+/-) granules" at peptide concentrations that are ~10-fold lower than the critical assembly limit. CATCH(+/-) granules are rapidly internalized by various adherent and suspended mammalian cell types. Over time, the active protein payload is delivered to the cytosol or other intracellular compartments. For example, CATCH-Cas9 co-delivered with a guide RNA can knockdown GFP signaling with comparable efficiency as "gold standard" lipofectamine vehicles, demonstrating CATCH-mediated protein delivery into the nucleus. Combining the CATCH(+) peptide with two different CATCH(-) fusion proteins provides granules with compositions governed by the molar ratio of the proteins in the mixture. These multi-protein CATCH(+/-) granules allow for co-delivery of the different protein payloads into the same cell, with user-defined control of the relative dose of each protein type. Importantly, this dual-component granule enables co-delivery of both an "effector" (i.e., a protein that confers therapeutic function) along with a "selector", where the latter can be used to enrich the sub-population of cells that have internalized the CATCH(+/-) granule. CATCH(+/-) granule internalization demonstrates greater delivery efficiency than other state-of-the-art methods, such as cell penetrating peptides or electroporation, without inducing cell death. We envision that this simple supramolecular nanomaterial approach will afford new opportunities for rapid and efficient cell engineering through intracellular protein delivery.

229f Desirable Formulations for Inhalable Particulate Immunotherapeutics Using Model Peg-Based Nanoparticle

Emma Sudduth, Catherine Fromen

The pulmonary microenvironment is the first site of infection for numerous pathogens, and as such, is comprised a host of specialized immune cells that are ripe for immunotherapeutic applications. However, determination of design parameters for inhaled nano-immunotherapies to target these cells has yet to be established. As a model platform, poly(ethylene glycol) diacrylate (PEGDA) nanoparticles (NPs) present highly beneficial characteristics for an inhalable formulation as they offer readily tunable physiochemical properties and have shown non-immunogenic effects in the lung. Herein, we characterize the aerosol and lung trafficking properties of model PEGDA NPs for inhalable immunotherapeutic applications through the modification of surface charge and formulation.

Both cationic and anionic formulations of PEGDA NPs were fabricated with similar hydrodynamic diameters around ~225 nm and low polydispersity indexes, through the addition of 10 wt% charge-establishing co-monomers. In liquid form, both formulations demonstrated robust nebulization through a commercial Aeroneb® device via non-significant morphological alterations post-nebulization, ascertained through Cryo-Scanning Electron Microscopy (Cryo-SEM) and ImageJ analysis. Dry-powder formulations of PEGDA NPs were also produced via lyophilization with tunable aerodynamic properties when dispersed through a Plastiape RS01 monodose inhaler. To assess phagocytosis capacity of PEGDA NPs using *in vitro* aerosol delivery, a 3D printed adapter was designed that connected the Aeroneb® device to air-liquid interface (ALI) transwell cultures of innate immune cells. As expected, cationic NPs demonstrated significant uptake at all dosing conditions based on electrostatic charges; however, NP uptake overall by these cells was greatly reduced at ALI, along with different phenotype responses. To corroborate findings from *in vitro* studies, murine studies of lung distribution following orotracheal delivery of cationic and anionic PEGDA NPs were performed using two different age groups. Cellular uptake in immune cell subpopulations, NP trafficking, and the cellular inflammatory profile were compared between the two groups for each of the particle formulations, suggesting distinct design rules for the different age groups. Thus, this work outlines an overall approach to formulate particulate-based inhalable therapeutics that can target specific immune cells in the pulmonary microenvironment for a broad range of inhaled immune engineering applications.

229g Macrophage Internalization of PEGDA Nanoparticles Drives Secretion of Pro-Survival Signals

Michael Trautmann-Rodriguez, Simone Sabnis, Catherine Fromen

Antigen presenting cells (APC), such as macrophages and dendritic cells, are phagocytic cells that internalize foreign objects and bridge innate and adaptive immune functions. These phagocytic cells naturally interact with engineered nanoparticles (NP) and receive a plethora of information during the phagocytic process that dictates the downstream response. Despite this knowledge, the full extent to which physiochemical cues from the NP drive changes to APC response is not fully understood, in large part due to the complexity of the phagocytic immune synapse. Prior work from the Fromen lab has shown that delivering model NPs comprised of poly(ethylene glycol) diacrylate (PEGDA)-based chemistries to primary macrophages enhances their survival, even in the absence of a therapeutic or stimulating cargo. Herein, we further investigate this phenomenon and study how this extended survival influences cell-cell communication. We first fabricate anionic PEGDA NPs ~200 nm in hydrodynamic diameter and evaluate cellular lifespan of two primary-like APC populations: murine bone marrow derived macrophages (BMM) and bone marrow derived dendritic cells (BMDC), both differentiated from C57BL/6 female mice. Expanding prior results, we demonstrate that a single dose of 100 μg/mL anionic PEGDA NPs significantly enhances *ex vivo* survival of both BMM and BMDCs; this effect is further modulated through repeated NP administration. Interestingly, we find that following NP-internalization, BMMs secrete an as-of-yet unidentified pro-survival signal that enhances survival of other BMM cells. Using either a single supernatant transfer or a non-contacting co-culture using transwells, exposure to supernatants of NP-dosed BMMs increases the survival of adjacent, NP-free BMMs to a significantly greater extent than those exposed to untreated BMM supernatants. Bioinformatics approaches evaluating both cellular RNAseq and supernatant mass spectrometry proteomic datasets have revealed a list of possible secreted factors accounting for this effect, lending important insight into the signaling cascade that is downstream of NP internalization. Fully characterizing the implications of phagocytosis on APCs such as BMMs builds important groundwork for a wide range of NP-based therapeutics and vaccines, where controlling the duration of survival and antigen presentation may offer an important, untapped engineering design parameter.

229h pH-Mediated Mucus Penetration of Zwitterionic Polydopamine-Modified Silica Nanoparticles

Yubin Ma, Bingbing Sun

The mucus lining on the mucosal surface works as a natural barrier against the invasion of pathogens. However, the existence of mucus hinders the trans-mucus delivery of medications, resulting in low administration efficiency and poor bioavailability. Zwitterionic polymers have emerged as promising trans-mucus nanocarriers due to their superior anti-fouling properties. However, the different behaviors caused by various pH environments have been observed in certain zwitterionic polymers. When delivered to various mucosal tissues, the pH-sensitive behaviors of zwitterionic polymers were noted due to a wide range pH micro-environment of mucus. The mucus layer is approximately neutral in airway (pH 7.1 ± 0.1) and eyes (pH ~7.26) of human. Along the gastrointestinal tract, the mucus pH increases from harsh acidic (pH 1-3) in stomach to slightly acidic or neutral in small intestine (pH 5.0-7.5). Additionally, the mucus pH of cervix ranges from slightly acidic to basic conditions (pH 5.4-8.2) with the menstrual cycle. Thus, the pH-sensitive zwitterionic materials could exhibit the desired mucus penetrability in the targeted mucus microenvironment that further determines their applicability as trans-mucus delivery vehicles. The systematic study on the effect of pH on the mucus penetrability of pH-sensitive zwitterionic NPs and the underlying mechanism is currently missing.

Herein, we prepared a library of zwitterionic polydopamine-modified silica nanoparticles (SiNPs-PDA). With the increase of the mass of PDA grafted on SiNPs, a library of SiNPs-PDA with different Si:DA ratios was prepared, and they were noted as SiNPs-PDA-11, SiNPs-PDA-12, and SiNPs-PDA-15, respectively. The Transmission electron microscopy (TEM) images and dynamic light scattering (DLS) analysis showed that pristine SiNPs were spherical-shaped with diameters of around 300 nm, and the coating became thicker and more compact with the increase of PDA. The Fourier transform infrared spectroscopy (FTIR), thermal gravimetric analysis (TGA), and acid orange analysis were used to determine the surface groups, which increased with the increase of DA:Si ratio. Besides, the ζ potentials of SiNPs-PDA were about 20 mV, 0 mV and -15 mV at pH values of 3.0, 5.6, and 7.0, respectively, and the isoelectric point (IEP) was around 5.6.

Using the library of pH-sensitive PDA-modified NPs, we further investigated their diffusion behavior in both *in vitro* and *in vivo* mucus models. The pH of reconstituted mucus was adjusted from acidic to neutral (pH= 3.0, 5.6, and 7.0) to mimic the pH microenvironment in different mucosal tissues, and multiple particle tracking (MPT) was used to determine the motility behavior of particles. The movement of SiNPs-PDA was displayed visually by representative trajectories and movies. It was shown that the SiNPs-PDA diffused freely at pH 5.6 and were restricted significantly as the mucus shifted to acidic (pH = 3.0) or neutral (pH = 7.0) condition. The ensemble mean squared displacement (<MSD>) was calculated to evaluate the diffusion ability of NPs. Under different modification densities, the SiNPs-PDA all exhibited the maximum <MSD> values in mucus with pH value of 5.6. It indicated that the balanced charges on SiNPs-PDA, even in a small amount, provided particles with good penetration ability. When mucus pH was changed to 3.0 or 7.0, the <MSD> values were reduced by different degrees. This reduction was augmented with the increase of PDA density, reaching 2.8 and 2.5 times for SiNP-PDA-15 at pH values of 3.0 and 7.0, respectively. It suggested that the pH-mediated mucus penetrability of SiNPs-PDA became more significant with the increase of unbalanced positive and negative charges on the particle surface. The logarithms of the effective diffusion coefficient (D_{eff}) and anomalous exponent (α) were calculated to further confirm the unhindered behavior of SiNPs-PDA at pH value of 5.6. Additionally, the mucosal tissues of mouse with different pH micro-environments were selected to further study the particle distribution *in vivo*. SiNPs-PDA-15 were diffused deeper into the intestinal villi (slightly acidic), and evenly distributed along the mucus layer. In contrast, SiNPs-PDA-15 were sparsely and discontinuously accumulated on the surface of the stomach (acidic) and airway (neutral), respectively. Together with the *in vitro* results, it further demonstrated that the SiNPs-PDA-15 tended to diffuse better in slightly acidic mucus but were trapped or removed in acidic and neutral mucus.

To understand the pH-mediated diffusion behavior of SiNPs-PDA in mucus, a series of biophysical characterizations were used to analyze this process. As the major protein in mucus, the mucin entangles and forms a dense fiber mesh against the diffusion of foreign particles. The particle-mucin interaction has been demonstrated as a key parameter in the filtering effect of mucus mesh, determining the mucus penetrability of particles. The interaction between SiNPs-PDA and mucin was systematically analyzed from the aspect of surface properties of NPs. For NPs in mucus, the highly hydrophobic surface could typically lead to a trapped state by mucin due to hydrophobic interaction. Thus, the hydrophobicity of SiNPs-PDA-15 at different pH values was determined. The pyrene assay was used for the hydrophobicity measurement, which indicated higher hydrophobicity of NPs at pH of 3.0 than that at pH of 5.6 and 7.0, respectively. Additionally, the Circular Dichroism (CD) spectrum showed that the α-helix in mucin was weakened and widened with the environmental pH values decreasing, caused by the change in stability of salt bridges between -COOH and -NH_2. With the unfolding of the glycosyl side chain, the hydrophobic regions were gradually exposed and the hydrophobicity of mucin was increased. Altogether, the higher hydrophobicity of both mucin and SiNP-PDA-15 might lead to the strong hydrophobic interaction at the pH of 3.0, while the hydrophobic interaction was weakened when the hydrophobicity of the mucin or NPs decreased at pH values of 5.6 and 7.0. Additionally, the interaction on the molecular level was investigated using isothermal titration calorimetry (ITC) analysis. The exothermic and entropy loss processes indicated that the interaction was dominated by the formation of non-covalent bonds, *e.g.*, electrostatic interactions and hydrogen bonds. Notably, the association constants (K_a) value at the pH value of 5.6 was around 7.7 and 1.3 times lower than that at pH values of 3.0 and 7.0, respectively, and the number of adsorption sites (n) at pH of 5.6 was smaller than that at pH value of 3.0 and 7.0. It suggested that the mucin-SiNPs-PDA-15 interaction was weaker at pH of 5.6 than that at 3.0 and 7.0, which was consistent with MPT analysis. Additionally, the hydrogen bond formation between SiNPs-PDA-15 and mucin was confirmed by the blue shift of the mucin peak in the amide I region after adsorption on particles using the FTIR analysis.

In this paper, we demonstrated the pH-mediated mucus penetration capability of zwitterionic polydopamine-coated silica nanoparticles. When the IEP of NPs was equal to the pH value of the mucus microenvironment, the NPs exhibited superior diffusional transportation in both *in vitro* and *in vivo* models. Further biophysical analysis found that the trans-mucus behavior of SiNPs-PDA was mediated by combined hydrophobic, electrostatic, and hydrogen bonding interactions between mucin and the particles. This study reveals the significance of matching the charge balance state with the mucus pH microenvironment for efficient trans-mucus delivery of zwitterionic NPs, which provides rational design strategies for applications of zwitterionic polymers in various mucus microenvironments.

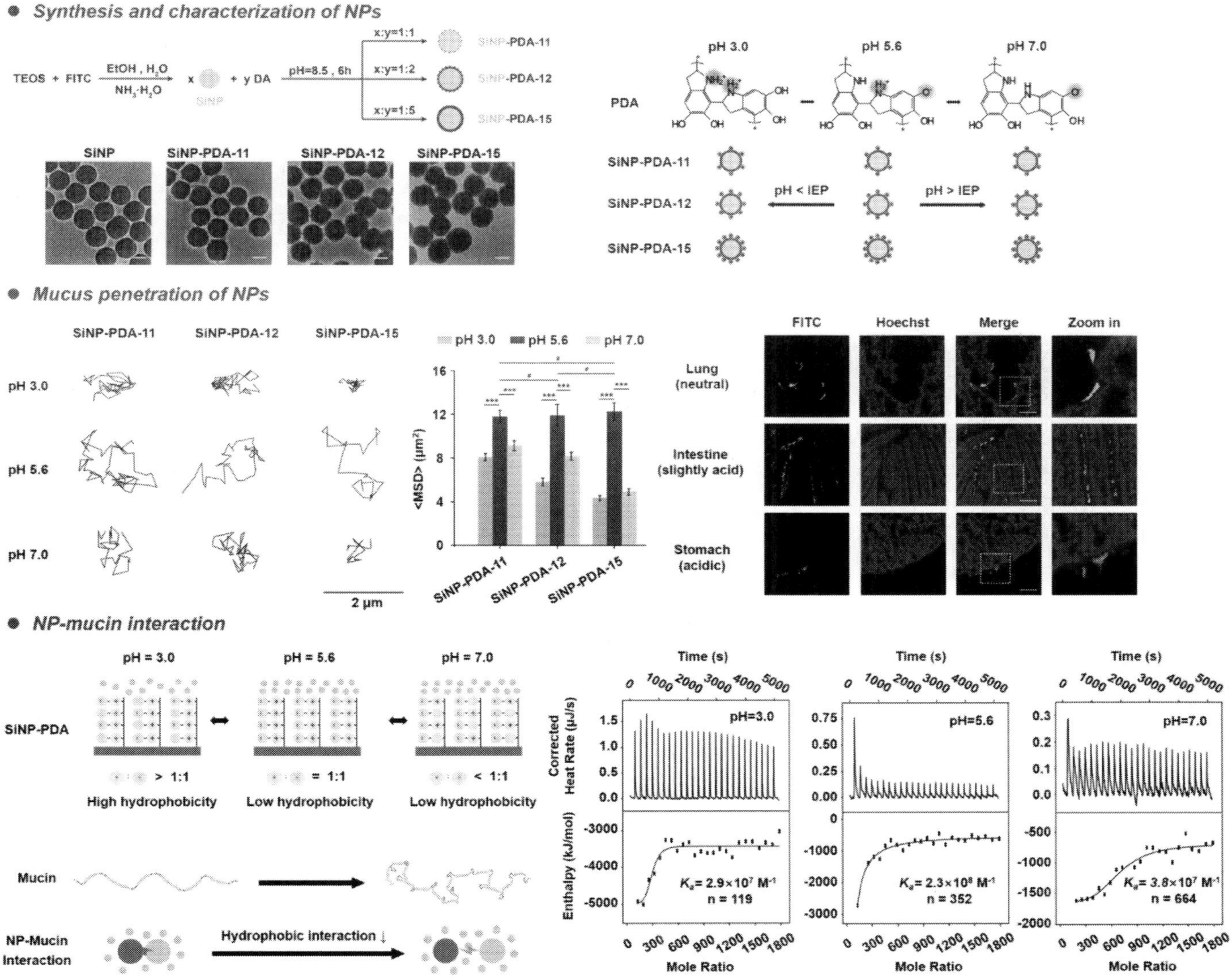
Synthesis and characterization of NPs
TEOS + FITC
EtOH, H_2O
$NH_3 \cdot H_2O$
x SiNP + y DA
pH=8.5, 6h
x:y=1:1
x:y=1:2
x:y=1:5
SiNP-PDA-11
SiNP-PDA-12
SiNP-PDA-15
SiNP
pH 3.0
pH 5.6
pH 7.0
PDA
pH < IEP
pH > IEP
Mucus penetration of NPs
pH 3.0
pH 5.6
pH 7.0
2 μm
<MSD> (μm²)
FITC
Hoechst
Merge
Zoom in
Lung (neutral)
Intestine (slightly acid)
Stomach (acidic)
NP-mucin interaction
pH = 3.0
pH = 5.6
pH = 7.0
SiNP-PDA
> 1:1
= 1:1
< 1:1
High hydrophobicity
Low hydrophobicity
Low hydrophobicity
Mucin
NP-Mucin Interaction
Hydrophobic interaction ↓
Time (s)
Corrected Heat Rate (μJ/s)
Enthalpy (kJ/mol)
Mole Ratio
pH=3.0
K_a = 2.9×10⁷ M⁻¹
n = 119
pH=5.6
K_a = 2.3×10⁸ M⁻¹
n = 352
pH=7.0
K_a = 3.8×10⁷ M⁻¹
n = 664

AUTHOR INDEX

Adams, Jaquesta 68
Adolphson, Alex 154
Akbulut, Mustafa 105
Al Mayyahi, Ahmed 125
Alamgir, Faisal M. 42
Albino, Ambar Velazquez 97, 98, 101, 130
Allam, Omar 35, 39, 52
Amama, Placidus 4, 125
Andersen, Mie 49
Ang, Mervin 70
Annabi, Nasim 1
Ao, Geyou 69, 86, 89
Arias, Tavio 68
Armani, Andrea M. 135
Armstrong, Alexander 130
Arnett, Natalie 83
Arral, Mariah 7, 56, 61, 92
Bagchi, Kushal 104
Baik, Chaekyung 31
Baker, Lane 80
Bazaid, Mohammed 34
Beidaghi, Majid 128
Bermudez-Berrios, Esteban 98, 101
Bhat, Bhargavi 105
Biswas, Pratim 26
Bonner, N'Quel 130
Bottenus, Nick 149
Broussard, Emily 47
Brunelli, Nicholas 87
Bunnett, Nigel 57
Chang, Hsueh-Chia 16
Chatzaki, Ioanna 144
Chaudhary, Namit 92
Chen, Zheng 40
Cheney, Zachary 100
Cheula, Raffaele 49
Chio, Linda 67
Cho, Junbeom 78
Cho, Sang-Joon 80
Choi, Je Yeon 31
Choi, Myunghoon 80
Chokshi, Varun 57
Chou, Che-Yu 114
Chowdhury, Sanchari 36
Chua, Nam-Hai 70
Chun, Hoje 48, 53
Chung, Minsub 54
Coon, Tiffany 92
Cunningham, Francis J. 67, 117
Daniels, Shelby 100
Datye, Abhaya 76
Davis, Virginia 128
Day, Nicole 149
Debele, Tilahun 94
Dennebouy, Lena 123
Deshmukh, Sanket 5
Dewey, Oliver 24
Dibiase, Beth 147
Dion, Gregory 94
Doerfler, Rose 92
Domayer, Madisen 154
Eichmann, Shannon 18
Esfahani, Milad 96
Eswaran, Neel 97
Everhart, Brian M. 4, 125
Fagan, Jeffrey 21
Farrokhi, Tannaz 119
Fei, Ling 47, 79
Figueroa, Anavi Benavides 82
Finale, Madeline 36
Foston, Marcus 26
Fox, Bethany 92
Fromen, Catherine 15, 155, 156
Fulmer, Michael 100
Gao, Fan 122
Gao, Xiaosi 45, 88
Gao, Ziang 88
Ge, Jun 124
Gholamirad, Farivash 120
Gkikas, Manos 119, 144
Gnanasekar, Pitchaimari 83
Gogos, Alexander 123
Goh, Natalie S. 117
Gomez-Bombarelli, Rafael 53, 77
Gong, Mingrui 82
Gong, Xun 30, 102
Good, Hayden 97
Goodwin, Andrew 2
Gordon, Kenneth 47, 79
Gualdron, Diego Gomez 4
Gunnels, Taylor 147
Guo, Yiyang 8, 145
Gupta, Ankur 149
Gupta, Yoshita 28
Hamade, Fatima 128
Han, Byungchan 48, 53

Han, Jeong Woo 41, 46
Harris, Michael T. 114
Hayman, Emily 117
He, Xiaoqing 125
Heller, Daniel 65
Hemmati, Shohreh 114
Heppe, Brandon 69, 89
Herrmann, Inge 123
Hilt, J. Zach 95
Hipp, Julie 29, 85
Hoare, Todd 151
Hong, Minjoon 48
Hubbard, Joshua 117
Hudalla, Greg 154
Hura, Gregory 117
Ichinose, Yota 24
Imhoff, Eric 97, 98, 101
Irvin, Glen 22
Iv, C. Wyatt Shields 149
Jabrayilov, Maharram 111
Jafari, Mina 62
Jagota, Anand 65
Jamali, Vida 18
Jana, Priya 87
Jang, Seung Soon 34, 35, 39, 52
Jiang, Peng 122
Jin, Shuo 45
Jin, Xiaojia 30, 66, 102
Jones, Christopher 75
Joo, Yong Lak 45, 88
Joseph, Karius 47
Jung, Hyunwook 50
Junnarkar, Jui 22
Kaemmer, Stefan 80
Kamat, Neha 147
Katnas, Eren 86
Kavadiya, Shalinee 26
Khabushev, Eldar 109
Khan, Faiz 87
Khomami, Bamin 103, 110
Khong, Duc-Thinh 70
Khoury, Joe 109
Kidambi, Piran 63
Kim, Bumjun 59
Kim, Dong Ha 35
Kim, Hyun Jin 54
Kim, Hyungjun 32
Kim, Mijin 65
Kim, Youngbi 46
Koman, Volodymyr 30, 66, 102
Komatsu, Natsumi 24, 71
Kono, Junichiro 24
Kotov, Nicholas 112
Kuehne, Matthias 30
Kumar, Prashant 112
Kumar, Sonu 16
Lachmayr, Kätchen K. 127
Landry, Markita 56, 61, 67, 68, 71, 117
Lavarone, Anthony 61
Ledesma, Francis 14, 67
Lee, Hyung Joo 31
Lee, Ilgeum 35
Lee, Jerry S. H. 135
Lee, Jin Gyun 149
Lee, Jong Suk 33
Lee, Joo-Youp 62
Lee, Sangmin 54
Lee, Won Bo 78
Lee, Youngmin 36
Leem, Esther 68
Leonard, Joshua 147
Lese, Ioana 123
Lewis, Parker 57
Li, Gao 115
Li, Lei 19, 81
Li, Muxiao 22
Li, Na 93
Lin, Yu-Ting 105
Liu, Jingyi 151
Liu, Qingsong 28, 29
Liu, Renjie 154
Liu, Shuhao 105
Liu, Sitong 130
Liu, Tsung Wei 4
Liu, Yu 124
Liu, Zhi 47, 79
Loesch-Fries, Sue 114
Loh, Suh In 70
Lopresti, Samuel 92
Lui, Alison 56, 61
Lundberg, Daniel James 102
Lustig, Steven 127
Lynn, David M. 3
Ma, Yubin 157
Mackintosh, Fred 18
Maeda, Yuki 137
Maestas-Olguin, Angelea 60
Malaney, Angela 92
Mandadapu, Kranthi K. 56
Maniya, Nalin 16
Margraf, Johannes T. 50
Maroudas, Dimitrios 25
Marras, Alexander 118
Marti-Arbona, Angel 109

Martinez, Payton ... 149
Maruyama, Benji ... 4
McCoy, John ... 36
McLeod, David C. ... 127
McReynolds, Brandon ... 36
Milano-Benitez, Mariana ... 103
Milliron, Delia ... 64
Mitchell, Duane A. ... 130
Mitrut, Roxana E. ... 147
Modi, Sujit ... 26
Mohammadi, Mohammad ... 140
Mokhtarnejad, Mahshid ... 103, 110
Monty-Bromer, Chelsea ... 100
Moore, Ginger ... 130
Moore, Le'Naedrea ... 79
Moradi, Yasaman ... 13, 135
Mostafa, Afnan ... 25
Mukherjee, Dibyendu ... 110
Mule, Dr. Aniket Sandip ... 106
Munoz, Enrique ... 37
Murali, Aishwarya ... 67
Nah, Yoonseo ... 35
Nakamura, Noriko ... 72, 132, 137, 141
Navarro, Nicole ... 68
Nealey, Paul F. ... 104
Neuer, Anna L. ... 123
Newby, Alexandra ... 92
Nguyen, Hao ... 43
Nguyen, Tu Huu ... 57
Nicholson, Marissa ... 95
Nishitani, Shoichi ... 67
Nißler, Robert ... 123
Ohta, Seiichi ... 72, 132, 137, 141
Okonkwo, Onochie ... 26
Ouassil, Nicholas ... 71
Pais, Michael A. ... 123
Pak, Chanho ... 31
Pandey, Aditya ... 88
Panthani, Matthew ... 111
Park, Cheol ... 109
Park, Minhee ... 78
Park, Yoonjee ... 94
Pasquali, Matteo ... 18, 22, 24, 82, 109
Paul, Pranto ... 95
Pawar, Shreya ... 90
Petersen, Daria ... 92
Pinals, Rebecca L. ... 61, 117
Pinkerton, Nathalie ... 57
Pollard, Rachel ... 57
Porter, Thomas K. ... 70
Powell, Matt ... 43
Preciado, Nahieli ... 10, 151
Prud'Homme, Robert K. ... 59
Pussepitiyalage, Chamath Vindula Basnayake ... 114
Qin, Zhaoxian ... 115
Raffaelle, Patrick R. ... 116
Raj, Ritu ... 149
Ramakrishnan, Subramanian ... 83
Ramasubramaniam, Ashwin ... 25
Ramos, Paolo ... 29, 85
Ranadive, Pinaki ... 87
Rao, Rahul ... 4
Razzaq, Atta Ur ... 96
Reed, Shane ... 83
Reuter, Karsten ... 50
Rezvan, Gelareh ... 120
Richards, Jeffrey ... 28, 29, 85
Rinaldi-Ramos, Carlos ... 97, 98, 101, 130
Romero, Cameron ... 47, 79
Rosenberg, Daniel J. ... 117
Ruskowski, Mary ... 144
Sabnis, Simone ... 156
Sadati, Monirosadat ... 120
Sadovsky, Yoel ... 92
Saha, Arnob Dipta ... 36
Saha, Sulay ... 26
Sahara, Yuto ... 132
Saju, Jolly ... 70
Sarker, Swagotom ... 125
Sarojam, Rajani ... 70
Scammell, Lyndsey R. ... 109
Schmidt, Brian ... 57
Schutte, Sophia ... 62
Senapati, Satyajyoti ... 16
Seo, Bumjoon ... 78
Shao, Yiqi ... 45, 88
Sharma, Kartikey ... 121
Shestopalov, Alexander ... 116
Shi, Yifeng ... 129
Siegrist, Theo ... 83
Sims, Christopher ... 21
Sinclair, John ... 16
Singh, Gajendra Pratap ... 70
Singh, Sanjay ... 83
Sloan, Arthur ... 22, 82
Solomon, Kevin ... 114
Son, Manki ... 66
Squire, Henry ... 61
Sreelatha, Sarangapani ... 70
Stark, Colin ... 151
Stege, Victoria ... 100
Stocker, Sina ... 50
Strano, Michael ... 30, 66, 70, 102
Subraveti, Sai Nikhil ... 59

Sudduth, Emma 15, 155
Sun, Bingbing 145, 157
Sun, Xiangcheng 20
Sun, Xiaoqi 68
Sun, Xue-Long 69, 89
Taheri-Qazvini, Nader 120
Tandabany, Dinadayalane 83
Tang, Zhao 18
Taylor, Lauren 24
Thome, Cooper P. 149
Thompson, Annette 19, 81
Tikhomirov, Grigory 129
Tiwari, Siddharth 88
Tokumaru, Mikito 72
Trafford, Mitchell 24
Trautmann-Rodriguez, Michael 156
Truong, Alice 144
Tsuchiya, Hiroki 141
Ucko, Amandine 30
Usie, Griffin 47
Vadthya, Raju 43
Vaidya, Akash J. 114
Veksler, Michael 112
Vinnedge, Lisa Privette 62
Visheratin, Alexander 112
Visheratina, Anastasiia 112
Voke, Elizabeth 56, 61
Walden, Mary 120
Wang, Jeffrey 117
Wang, Song 70
Wang, Yuhuang 65
Wang, Yunshan 140
Waterkotte, Thomas 94
Wehmeyer, Geoff 24
Wei, Shuya 43
Wetzel, Eric D. 127
Whitehead, Kathryn 56, 61, 92
Winter, Jessica 87
Woods, Mackenzie 128
Wu, Jiachen 122
Yanagi, Kazuhiro 24
Yang, Fan 19, 81
Yerneni, Saigopalakrishna 92
Yomogida, Yohei 24
Yu, Bo 6, 130
Zhao, Meng-Qiang 126
Zhao, Xiaoxiao 108
Zheng, Ming 65
Zheng, Zhiwei 94
Zhou, Fei 51
Zhou, Hao 26
Zimmermann, Monika 123